DRIVING FORCES IN PHYSICAL, BIOLOGICAL AND SOCIO-ECONOMIC PHENOMENA

A Network Science Investigation of Social Bonds and Interactions

In recent years network science has become a dynamic and promising discipline; here it is extended to explore social and historical phenomena. While we experience social interactions every day, there is little quantitative knowledge on them. Instead, we are often tempted to resort to fanciful explanations to account for social trends. For example, it has been argued that the decrease in suicide rates in America in the 1990s should be attributed to greater consumption of anti-depressants. However, further examination revealed that US counties where suicides rates have fallen the most are those with a high proportion of Hispanic immigrants, who are known to have low suicide rates. More generally, exogenous and endogenous interactions are often the key to understanding social phenomena and unravelling historical mysteries.

This book begins by explaining how it is possible to bridge the gap between physics and sociology by exploring how network theory can apply to both. It then examines the macro- and micro-interactions in societies. The chapters are largely self-contained, allowing readers to easily access and understand the sections of most interest to them. This multidisciplinary book will be fascinating to all physicists who have an interest in the human sciences, and it will provide an alternative perspective to graduate students and researchers in sociology and econophysics.

BERTRAND M. ROEHNER is a Professor at the Institute for Theoretical and High Energy Physics at the University of Paris, France. He has written several books on econophysics, including *Patterns of Speculation* (Cambridge University Press, 2002) and *Pattern and Repertoire in History* (Harvard University Press, 2002).

DRIVING FORCES IN PHYSICAL, BIOLOGICAL AND SOCIO-ECONOMIC PHENOMENA

A Network Science Investigation of Social Bonds and Interactions

BERTRAND M. ROEHNER

University of Paris

CAMBRIDGE
UNIVERSITY PRESS

CAMBRIDGE UNIVERSITY PRESS
Cambridge, New York, Melbourne, Madrid, Cape Town, Singapore, São Paulo

Cambridge University Press
The Edinburgh Building, Cambridge CB2 8RU, UK

Published in the United States of America by Cambridge University Press, New York

www.cambridge.org
Information on this title: www.cambridge.org/9780521859103

© B. M. Roehner 2007

First published 2007

Printed in the United Kingdom at the University Press, Cambridge

A catalog record for this publication is available from the British Library

ISBN 978-0-521-85910-3 hardback

Facts are stubborn things; and whatever may be our wishes, our inclinations, or the dictates of our passions, they cannot alter the state of facts and evidence.

John Adams, December 1770

We continued our systematic survey of the edge of the sodden portion of the moor, and soon our perseverance was gloriously rewarded. Right across the lower part of the bog lay a miry path. Holmes gave a cry of delight as he approached it. "Here is Herr Heidegger, sure enough! My reasoning seems to have been pretty sound, Watson."

Sir Arthur Conan Doyle, "The Adventure of the Priory School" (1905)

Contents

Preface

The purpose of this book is to explore the similarities as well as the differences between natural and social phenomena. While several chapters are devoted to the second objective, the main message of the book is to show that the experimental methodology which has been used so successfully in the natural sciences can also be applied to social phenomena.

Basically, one can distinguish two main approaches in the social sciences, the anthropomorphic perspective and the system theory approach. Throughout this book we try the second of these options. In order to illustrate them, let us consider the phenomenon of suicide. In the anthropomorphic perspective one would try to establish connections between suicide and psychiatric disorders. In the system theory approach one tries to formulate the problem of suicide in a more general framework in which the interactions between the elements composing the system play the key role.[1] This approach was pioneered by the sociologist Emile Durkheim at the end of the nineteenth century. He was able to show that there is a clear relationship between the likelihood of suicide and the strength of the bonds which link an individual to the rest of the society. The main challenge (and prerequisite) in the application of the network science approach is to be able to estimate the strength of the interactions between the elements of the system. This is also the main stumbling-block because both system theory and network science are mathematical frameworks which provide useful modeling tools but do not tell us how to investigate complex, interconnected networks and how to measure the interactions between their elements. Network theory will not show us how complex problems can be simplified, nor will it tell us anything about experimental procedures aimed at estimating the strength of bonds. In short,

---

[1] As a consequence this approach is not restricted to human societies but may be extended (at least if the interactions are similar) to biological or physical systems. Such a perspective was developed by system theory in the 1960s and 1970s and was revisited and revived by network science in the late 1990s and early 2000s.

although we need some guidance for the exploration of this new and uncharted territory, we can hardly find it in network science alone.

Is there another field which can guide us? Physics and chemistry have been trying to understand systems of interacting entities for over three centuries. In situations where one is confronted with an opaque maze of interconnected and multifaceted systems, chemistry and physics have developed methods which enable us to probe one kind of interaction at a time, for instance by exploring the phenomenon at different scales in space and time. This is why in this book we often use such methods as a source of inspiration, a reservoir of ideas and solutions. Parallels between physical and social phenomena could be seen merely as analogies; that would be a narrow perspective, however. In our view these parallels reflect real underlying similarities in terms of interactions and network structures. Naturally, we understand and accept the fact that this is a point on which opinions may differ. The ultimate test is whether this approach can help us to build up a better understanding of social phenomena.

To carry out this program one needs to take a fresh look not only at social phenomena but also at physics. Indeed, if parallels with social phenomena are to be found, it is certainly not at the level of fields such as general relativity or string theory that they can be discovered. The parts of physics which seem the most promising are for instance the rules which describe the mixture and miscibility of liquids, the solubility of gases in liquids, phase transitions between different allotropic structures, and other similar issues. Such questions were actively investigated in the nineteenth and early twentieth centuries but are somewhat neglected nowadays. This is why our exploration is also a journey in some forgotten corners of physics. In a general way, our sources of inspiration are not so much the mathematical theories of the physical phenomena under consideration, but rather their understanding in terms of molecular mechanisms and interactions.

During past decades social scientists have had increasing recourse to mathematical tools whether in the form of statistical tests, computer simulations or mathematical models. This may give the impression that these models have become more "scientific" and in a sense closer to the natural sciences. However, this observation is called into question by closer examination.

The first and somewhat surprising observation is that a similar evolution is under way in physics as well. As an illustration, one can mention the fact that between 1900 and 2000 the proportion of purely theoretical papers in *Physical Review*, one of the main physics journals, has increased from 10% to 55%.[2] Does

---

[2] By "purely theoretical" we mean papers that do not make contact with actual data. At the same time the proportion of experimental papers has decreased from 85% to about 35%; a third but rather small category of papers is represented by theoretical papers which make contact with experimental results. If this evolution continues, by 2050 experimental and purely theoretical papers will represent 15% and 80% respectively. (For more details see Roehner 2004, p. 327.)

this increased mathematization mean that physics has become more scientific and more productive than it was in the early twentieth century, a time marked by the emergence of statistical physics, quantum mechanics and general relativity? For a solution of this paradox, one should recall that "more mathematical" does not necessarily mean "more scientific" and vice versa. As an illustration, let us consider Galileo's celebrated experiment of a ball rolling down a ramp, which is viewed as the starting point of modern physics. This experiment can be set up and interpreted without any mathematical knowledge beyond basic arithmetic. The law governing falling bodies derived from the experiment was stated by Galileo in the following terms: "We always found that the spaces traversed by the ball were to each other as the squares of the times and this was true for all inclinations and for all balls."[3] What was really new in this experiment was that (i) it concerned a *simple* phenomenon in the sense that its design made friction and other side effects almost negligible, and (ii) the measurements were carried out with high accuracy and repeated a great number of times, "a full hundred times" writes Galileo. We will see in a subsequent chapter why the number of repetitions is a crucial element in the battle against noise.

These observations suggest three key conditions for a scientific observation: (i) to investigate one phenomenon at a time; (ii) to make sure that the system under consideration is closed or, if it is not, that all exogenous factors are duly taken into account; (iii) to carry out the measurements with utmost accuracy. These rules are of course well known, but paying lip service to them is not sufficient; in truth, in the social sciences they are rarely applied. The first rule is probably the most difficult to comply with because most questions in the social sciences are multifaceted issues which have no real meaning unless one is able to disentangle the various mechanisms which are involved. The second rule also has broad implications. Historians, especially when they write the history of their own country, have a marked tendency to forget, discard or belittle exogenous influences. In writing a history of the administration of justice in France it probably makes sense to neglect exogenous factors, but writing a history of French labor unions in the twentieth century without due recognition of American or Soviet interferences would make little sense. Similarly, a history of the French Revolution leaving out British influence and interventions would not be very realistic. And yet, most of the works on these topics devote at best a few lines to such exogenous factors. The question of hidden exogenous influence will be considered in more detail in a subsequent chapter.

One explanation of the gap between physics and the social sciences can be found in their different traditions regarding the question of how to make

---

[3] *Dialogue Concerning Two New Sciences*, Macmillan (1914, p. 178).

measurements.[4] Physicists believe in laws which have a permanent and universal validity. They are prepared to devote years or even decades to establishing, validating and confirming such laws. As an example, one can mention the experimental tests of the theory of general relativity. They began in 1919 when two British teams led by Eddington took advantage of a solar eclipse to measure the deviation of a beam of light emitted by a star when it travels through the the gravitational field of the Sun. Separate observations were made with three different instruments. Einstein's prediction was confirmed by two of the three sets of measurements but not by the third; thus, the question could not be considered as completely settled.[5] During subsequent decades, astronomers took advantage of each total eclipse to repeat the measurement and obtained rather conflicting results. After World War II similar observations became possible with radiotelescopes, which had the advantage of being independent of the occurrence of eclipses. Each measurement had only a low precision, but by repeating them for hundreds of stars it was possible to improve the accuracy. In the 1990s very accurate atomic clocks were put aboard spacecraft and provided other tests of general relativity apart from the deviation of a beam of light. Finally, in the early 2000s several huge facilities were set up in different countries in order to track the gravitational waves which are predicted by Einstein's theory but have not yet been observed. In short, for almost a century the predictions of general relativity have been tested and retested unremittingly.

To the contrary, social scientists do not expect to find laws of permanent and universal validity. As a matter of fact, the very idea that there can be universal laws is opposed by many social scientists. As a result, only limited time and effort are devoted to making accurate measurements. Why should one bother to improve the accuracy of a law if one knows in advance that its validity will not outlast a couple of decades?

What is the position of social scientists on the question of bonds and interactions which is the topic of this book? First of all, one can remark that the situation is not at all the same as in physics. As we know, the existence of molecules and the role of intermolecular bonds remained uncertain until the early twentieth century. It was only after World War I that the strength of molecular interactions could be measured reliably. In contrast there can be little doubt that interpersonal relations exist and play a role in social phenomena. So far, however, we are unable to evaluate their strength. Why is it so difficult to gauge the strength of interpersonal bonds? In a subsequent chapter we argue that it is the level of "noise" which is the main obstacle. Every person has many different roles, being

---

[4] This is what Harvard sociologist Stanley Lieberson calls "making the number count" (for more details see Lieberson 1980, 1985).

[5] Newtonian mechanics also predicted a deviation but of a different magnitude; therefore it is essential to measure the deviation with high accuracy.

simultaneously a husband or wife, a citizen, a consumer, a member of a religious community and so on and so forth. When one tries to focus on only one of these facets, the other interactions are nevertheless present and act as perturbing factors which, for the sake of brevity, we summarize under the term "noise". In the study of social phenomena one of the main challenges is to improve the signal to noise ratio. Two subsequent chapters are devoted to this question.

The present study contains few theoretical models; this is intentional and deserves an explanation. One should keep in mind that there is a fundamental difference between model building in physics and model building in the social sciences. In physics there are several fundamental principles, such as Newton's laws, the first, second and third laws of thermodynamics, the conservation of energy, of momentum, of angular momentum, etc. Any model must be consistent with these principles, which means that model building is highly constrained and circumscribed. As a result, any model which is also consistent with experimental evidence will be a meaningful model. In the social sciences there are almost no basic principles. Consequently, it is possible to build a great variety of models. As an illustration, one can mention that there are at least three different strands of stock market models. Some are based on a competition between informed investors and so-called "noise traders"; others are based on game theory; a third group relies on extreme value theory; one could mention several other varieties. Needless to say, all these models are able to explain the "stylized empirical behavior" of share prices. In this book we rather rely on what can be called *regularity models*. The notion can be explained by way of a simple illustration. Suppose that I wish to predict the rising time of the Sun. If on Monday I see it rise at 6:00, on Tuesday at 6:05, on Wednesday at 6:10, it will be natural to expect it to rise at 6:15 on Thursday and at 6:20 on Friday. This example may seem trivial but it can easily be complicated. Suppose that during the weekend I leave Paris to visit a friend in Brest, 600 kilometers to the west; once in Brest, the prediction for sunrise time will prove incorrect by at least 20 minutes, which suggests that the initial model was too rudimentary. Furthermore, the model will also prove incorrect if one tries to use it on a time scale of several months. Thus, step by step, one will be able to improve the initial model. Many fields of physics in which there are no simple laws are organized in this way. For instance, Bernoulli's theorem serves as a *regularity model* in hydrodynamics in the sense that it "explains" several effects (e.g. the Venturi effect, the lift of an airfoil, the velocity of water exiting through a hole at the bottom of a tank, etc.), albeit with rather poor accuracy. As the standard Bernoulli formula does not take into account viscosity or turbulence it can serve only as a rough approximation. In each case, if one wants an accurate description, the Bernoulli model must be improved in several respects.

Writing this book has been an exhilarating journey, in the course of which we have roamed through several social phenomena, various historical periods and many databases. In some places, we may have erred. That is almost inevitable if one considers the diversity of the data that needed to be processed. We welcome in advance any notification of possible errors or omissions. Few readers are likely to read this book from cover to cover. This is why most of the chapters have been written in a way which makes them readable independently of the others;[6] in particular, some basic definitions and arguments have been purposely repeated in different places whenever they were needed in order to make each chapter almost self-contained.

It is a pleasure at this point to thank the many people who provided encouragement, help, support and advice. Throughout this research, the quest for various sorts of data has been a permanent concern. Naturally, the Internet was of great help; yet many of the data that we needed were not available online but were kindly sent to us by helpful researchers working in statistical institutes or documentation centers. Many thanks to Maureen Annets (Commonwealth War Graves Commission), Dan Bernhardson (National Board of Health and Welfare, Sweden), Charlotte Björkenstam (Socialstyrelsen, Stockholm), Anita Brock (Mortality Statistics, UK), Birgitta Chisena (Statistics Office, Sweden), Adela Clayton (Australian War Memorial), Eddy Dufournont (INALCO, Paris), Christine Hauchecorne and Gilbert Chambon (Interlibrary loan, University of Paris 6–7), Annick Horiuchi (Department of Japanese Studies, University of Paris 7), Dorthe Larsen (Statistics Office, Denmark), Wenhui Li (Bureau of Vital Statistics, New York City Department of Health), Frida Lundgren (National Board of Health and Welfare, Sweden), Ron Orwin (British Commonwealth Occupation Force Executive Council of Australia), Gunvor Stevold (Statistics Office, Norway), Kenneth Schlessinger (National Archives and Records Administration, United States), Jean-Luc Stroobant and Isabelle Masson (Statistical and Economic Information Office, Belgium), and Erwin K. Wüest (Statistics Office of Switzerland).

During the time I was writing this book I had interesting and stimulating discussions with several of my colleagues, and in particular with Masanao Aoki (University of California at Los Angeles), Marcel Ausloos (University of Liège), Belal Baaquie (University of Singapore), Doyne Farmer (Santa Fe Institute), Michael Hechter (University of Washington), Gérard Jorland (Ecole des Hautes Etudes en Sciences Sociales, Paris), Taisei Kaizoji (International Christian University, Tokyo), Stanley Lieberson (Harvard University), Thomas Lux (University of Kiel), Bruce Mizrach (Rutgers University), Denise Pumain (University of Paris

---

[6] Naturally, there are in fact many connections between the different chapters, particularly between the chapters in Part I and those in Parts II and III; Part I sets the methodological guidelines which will be put to use in the two following parts.

1 Panthéon), Peter Richmond (Trinity College, Dublin), Alberto Sicilia University of Paris 6), Dietrich Stauffer (University of Cologne), Hideki Takayasu (Sony Computer Science Laboratories, Tokyo), Peter Turchin (University of Connecticut) and Ainslie Yuen (Goldman Sachs, New York). I would like to mention particularly my collaborations with Liguori Jego (Ecole Polytechnique), Sergei Maslov (Brookhaven National Laboratory) and Didier Sornette (Swiss Federal Institute of Technology, Zurich, and Centre National de la Recherche Scientifique). Through his commitment to this project Liguori helped me to bridge the gap between physical and social science conceptions; he also read substantial sections of the manuscript and provided clear-sighted remarks and suggestions; I thank him very much for his interest and cheerfulness. At several crucial junctures, Sergei was able to come up with hints which allowed significant progress and I am deeply grateful for his invigorating guidance. My collaboration and companionship with Didier started more than ten years ago in the early days of econophysics; based on shared views and common interests, it continued with the publication of several joint papers; many thanks to him for his open-mindedness, enthusiasm, insight and good humor.

Colleagues at my laboratory in Paris have been of great assistance on many occasions. I am particularly grateful to Olivier Babelon, Laurent Baulieu, Matteo Cacciari, Bernard Diu, Benoît Douçot, Jean Letessier, Marie-Christine Lévy, Annie Richard, Valérie Sabouraud, Claude Viallet and Bernard Zuber.

Through its physics publishing director, Simon Capelin, Cambridge University Press played a pioneering role in the development of econophysics. As is the case for any new field, the first studies and publications were essential, and the collaboration between their authors and the publisher was instrumental in shaping and delimiting the new field.[7] Since the early days of the mid 1990s the field has mellowed and its perspective has broadened. If the "physical approach" indeed opens a new frontier in the social sciences, there is no reason to confine it to financial analysis or to economic issues.[8] It is this challenge which led to the writing of the present book. Once again, as on former occasions, the contacts I have had with Simon have been a source of inspiration and I am most appreciative of his insight, perceptiveness and vision.

It is perhaps not surprising that the author of a book about social links is particularly aware of his debt to many unknown people whose activity and efforts permitted the fulfillment of this work. Reliable postal services, efficient electric utilities, convenient public transportation facilities and many other amenities were essential in the completion of this project. I express my deep gratitude to all these

---

[7] It can be recalled that the first book to appear in this field was *An Introduction to Econophysics* by Rosario Mantegna and Eugene Stanley; it was published by Cambridge University Press in 2000.

[8] In an earlier publication (Roehner 2002, Chapter 3) we made the point that most economic problems ultimately are conditioned by sociological phenomena.

people and would like to extend special thanks to the drivers of the subway line number 10 (Gare d'Austerlitz–Pont de Sèvres) which I took several times a week to go from my university to the National Foundation for Political Science (FNSP, Paris).

This book is dedicated to my wife, Brigitte, and to my son, Sylvain, whose cheerful encouragement and stimulating support were invaluable.

Part I

Bridging the gap between physics and the
social sciences

The emergence of two new fields, the science of chaos and the science of networks, has changed the way in which we look at physical and social systems. From the first we learned that simple (in the sense of having few degrees of freedom) physical systems can undergo chaotic motions and display intricate trajectories. The double pendulum is one of the simplest systems of this kind. Initiated by Benoît Mandelbrot (1975) and Robert May (1976), the science of chaotic systems and fractals has already produced substantial achievements. This book relies only occasionally on the analysis of chaos; in contrast it relies heavily on the ideas of network science. Although it can be traced back to system theory which flourished in the 1960s and 1970s, network science really emerged in the late 1990s through the works of people such as Albert-László Barabási (2002), Sergei Maslov (2002), Steven Strogatz (2003) or Duncan Watts (2003). It has been instrumental in convincing us that what really matters in a system is its nodes, its links and their respective weights. Seen in this perspective, the real nature of the system, whether of physical, biological or social nature, is of little relevance.

But looking at physical and social systems in an abstract, purely structural way takes away much of their substance. The real challenge is to do real physics and real sociology in the framework of network theory. This is what we call "bridging the gap". In the five chapters which compose this first part we analyze the implications of a perspective based on network science without losing contact with real systems. We consider the problems of measuring the strength of bonds, of reducing the level of noise in social systems; we discuss the differences between equilibrium and non-equilibrium phenomena. Finally, by way of specific examples, we emphasize that the question of data reliability has so far received too little attention in the social sciences.

1

Probing bonds

The fact that interactions and bonds play an essential role in physical phenomena can hardly be disputed. Their role in the social world is no less important. The parallel between these two classes of phenomena can be illustrated by two key episodes: the formation of stars and the emergence of agrarian-based societies.

- In the early universe there were molecular clouds composed of low density hydrogen and helium gas. According to current conceptions of star formation, regions where density was high enough (as a result of random fluctuations) became gravitationally unstable and began to collapse. The gravitational collapse released a lot of heat and energy in the same way as a stretched spring releases its potential energy as it resumes its equilibrium length. Part of the energy was radiated but the remainder increased the temperature of the core until fusion ignition occurred. This marked the birth of a new star. If its mass was greater than five solar masses it had a fairly short life time. After several million years its core began to contract and as it shrank it grew hotter; this triggered a new series of nuclear reactions leading to the formation of heavier elements up to iron. Eventually, through a mechanism which is not yet clearly understood, the star exploded. It is known that elements heavier than iron were formed during this supernova explosion. Through its disintegration, all elements contained in the star were released into space to serve as raw material for new stars, planets and living creatures.
- The neolithic revolution occurred only about ten thousand years ago but its successive steps are not so well known as those in the evolution of stars. Before the neolithic revolution there were low density populations living a largely nomadic life based on hunting and gathering. For some reason, a higher than average density of population developed in some places, therefore limiting the territory available to nomadism. Gradually permanent or semi-permanent settlements appeared, and gathering techniques became more intensive, eventually leading to crop cultivation. As the improved techniques could sustain more important population densities, population concentration continued, leading to the formation of towns and cities and to the emergence of techniques which represent important landmarks in the evolution of mankind: storing grains, making pottery, developing written languages, and introducing political and religious forms of social organization. The first cities may have lasted a couple of centuries but eventually

wars, diseases or internal factors made them crumble and collapse. While some of the newly developed techniques may have been lost, it seems reasonable to assume that most of them were disseminated and recycled in new settlements.

In short, from a network perspective, the two scenarios comprised the following steps. (i) Contraction brings about more interactions. (ii) Greater interaction leads to the formation of more complex entities. (iii) Disintegration eventually produces a spatial dissemination of the new entities. (iv) These new entities are reused as building blocks in subsequent processes. If this parallel does not appear completely convincing it is probably due to the following reason. Whereas we are familiar with the concepts of gravitational and nuclear attractions which account for the evolution of stars, we have no specific concepts for characterizing and differentiating the interactions in a society of hunters on the one hand and in an urban society on the other hand. Of course, we can explain and outline the differences but we have no means for describing them quantitatively. For a society there are no notions of temperature, heat, energy or entropy. We do not know how to characterize the interactions between wheat farmers, soldiers and priests in Ancient Egypt for instance.

In the same line of thought it is tempting, especially for a physicist, to mention a third kind of episode, namely cultural and scientific revolutions. Among several possible cases, we select the birth of quantum mechanics in 1925–8. We will see that it is not unlike the birth and explosion of a hot supergiant star. In the previous three decades, a wealth of experimental results had been produced which had received no consistent and comprehensive explanation. One can mention for instance the discovery of radioactivity, the emission spectra of atoms, the emission of α-particles by nuclei, the absorption coefficients of X-rays, the extraction of electrons from a metal by an electric field, and so on. In spite of a number of scattered phenomenological attempts there was still no comprehensive theory in the early 1920s. Then, within a few years, not one but several frameworks were proposed which eventually proved to be equivalent. One may wonder what was the role of social networks in this revolution.

As the group of physicists engaged in the exploration of the atomic world was a small community they often met and knew one another fairly well; thus there were links between Niels Bohr, Albert Einstein, Max Planck and Ernest Rutherford, to mention just a few of them. This community had a crucial function as a scientific filter in the sense that new theories were discussed, tested against evidence from new experiments, and (most often) were found wanting. This was a kind of potential barrier which prevented the adoption of unsatisfactory, makeshift theories. That was the normal mode of operation of the system, but between 1925 and 1928 it worked with much greater speed and efficiency. Berlin, Cambridge, Copenhagen, Göttingen, Leiden and Munich had for years been magnets attracting young talent; after 1925 it seems that Göttingen became the center of this network. Is it not revealing that in 1926–7 Arthur Compton, Paul Dirac, Linus Pauling, John von Neuman, Robert Oppenheimer, Edward Teller and Eugene Wigner, none of whom was German, visited Göttingen, staying there for several weeks or months and meeting one another (Rival 1995)? They were young and were not the main actors in this revolution, but they took part in it. The new ideas were actively discussed during many informal gatherings either at Max Born's home or at one of the inviting inns that could be found in the countryside surrounding Göttingen. In 1928, when the supernova exploded, the Göttingen researchers were scattered far and wide. In subsequent years, they would apply the new theoretical tools they had mastered to various subjects from chemistry to nuclear physics, astrophysics and several other fields.

Many other episodes similar to the quantum revolution can be found in the history of science. The process seems to be always the same. (i) Accumulation of precise and unexplained experimental results. (ii) This body of evidence acts as a converging lens through which the efforts of many researchers become focused on the same objective. (iii) This concentration of efforts eventually leads to a new theoretical framework. The revolution occurs in a place where the concentration process has been particularly intense but often another (more or less equivalent) solution is proposed almost simultaneously by another group of researchers. (iv) After the breakthrough the new tools are exported to other fields where they are employed with success. Describing these mechanisms quantitatively appears even more difficult than for the neolithic revolution. Nonetheless, it is possible to define objective criteria; for instance, a possible way of describing the links between different groups of physicists would be to use the academic genealogy method introduced by Sooyoung Chang (2003), an approach which is based on the links between thesis advisers and their graduate students.

In this chapter and indeed in the whole book we will try to develop a number of tools which can help us to measure social interactions. In order to be in the most favorable position we will mostly focus on phenomena for which vast statistical databases are available; this is for instance the case for commodity markets, suicide rates or international relations. Because in physics we have elaborate knowledge of how to measure bonds and interactions, it is natural to use those parts of this knowledge which can be useful for our purpose.

The present chapter is organized as follows. First, as a case in point, we describe in some detail the experiment which led Rutherford to the discovery of the nucleus. Apart from its historical importance, this example will show how important it is that the probe (in this case α-particles) is well adapted to the size of the entities that one wants to identify. After this example, we list and describe other means that can be used to gauge interactions. Finally, in the last section we briefly review some methods such as correlation analysis which are already used in the social sciences; we explain why they are not satisfactory for the purpose of measuring interactions and we examine how they can be improved.

1.1 The Rutherford experiment

The Rutherford experiment can be considered as a particularly successful illustration of a general strategy which can be summarized as follows. The system under consideration is subjected to a controlled exogenous change and from the way it responds to this change one tries to derive the form, range and strength of the interaction.

In many descriptions of the Rutherford experiment it is claimed that its main novelty and achievement was the fact that some of the incident particles were scattered backwards. In fact, experiments in which the incident particles were highly deflected to the point of re-emerging from the same side of the target had been performed several years before the Rutherford experiment. What was really new was that Rutherford and his collaborators had good reasons to think that in their experiment the reflected particles had experienced single (and not multiple) collisions. Their argument relied on the observation that most particles were little deflected; as a matter of fact, 99.990% of the particles experienced very small deflections. Thus, paradoxically, the particles which were not deflected were as important as those which bounced backward. After a short description of the experiment we explain why its interpretation was by no means straightforward. Finally, we discuss possible implications for our topic.

As we know, Rutherford used α-particles to explore the structure of gold atoms. It is important to realize that several other probes were available at the time, particularly cathode rays, X-rays and β-rays.[1] Table 1.1 shows that the α-particles were the only probe capable of exploring the nucleus. The wavelengths of the other probes were too large. Before discussing further implications let us describe the experiment. The α-particles were directed normally onto a very thin sheet of gold (Fig. 1.1). A fluorescent screen (made up of zinc sulfide) which produced tiny flashes of light when hit by α-particles was used to detect the scattered particles. The number of scintillations per minute and per square millimeter of the screen was counted by means of a low power microscope. There were three kinds of trajectories. (i) A fraction of the α-particles was absorbed by

---

[1] Let us recall the definitions of these expressions in modern terms. Cathode rays are streams of energetic electrons, X-rays are energetic electromagnetic waves, β^- rays are streams of electrons, β^+ are streams of positrons. α-rays are helium nuclei composed of two protons and two neutrons. At the end of the nineteenth century the word "rays" was used in a fairly vague way because the exact nature of these rays was unknown. In modern terminology, the term "ray" has been kept only for X-rays which are indeed a form of light radiation.

Table 1.1 *Probes available around 1909 for exploring the structure of the atom*

Name of probe	Nature of probe	Charge	Wavelength (10^{-14} meter)
Cathode rays	electrons	−	1,000
X-rays	photons	0	1,000
β-rays	electrons or positrons	− +	10
α-rays	2 protons + 2 neutrons	++	0.1

Notes: The diameter of a gold nucleus is about $d = 10^{-14}$ meter. Only probes with a wavelength shorter than d will be able to detect the nucleus. This criterion generalizes the rule well known in optics which says that, whatever its magnification power, a microscope will not permit the observation of details smaller than the wavelength of visible light (i.e. 4×10^{-6} m). The wavelength of cathode rays is determined by the voltage between cathode and anode; to get an order of magnitude we consider the wavelength of modern electron microscopes. The energies of the β- and α-rays available at the time were of the same order of magnitude, about a few Mev. For a particle of kinetic energy E whose velocity v is less than 40% of the speed of light, one can apply the following non-relativistic formulas (the difference with respect to relativistic formulas is less than 10%): $v = \sqrt{2E/m}$, $\lambda = 2h/mv = h\sqrt{2/Em}$ where h is Planck's constant, m the mass of the particle, and λ its associated wavelength. As the mass of an α-particle is about 8,000 times the mass of an electron, its wavelength will be $\sqrt{8,000} \sim 90$ times smaller than the wavelength of a β-particle of same energy.

the target. (ii) Another fraction was transmitted through the gold sheet without being substantially deviated. (iii) A small fraction (about 0.01%) of the scattered particles experienced a deflection of more than 90 degrees which means that they emerged from the same side of the plate as that on which they hit the target. The fact that scattering of more than 90 degrees had already been observed for β-particles is clearly stated at the beginning of the paper by Geiger and Marsden (1909) who were Rutherford's collaborators: "When β-particles fall on a plate they are scattered inside the material to such an extent that they emerge again at the same side of the plate. For α-particles, a similar effect has not previously been observed." What made the reflections of the α-particles important was the fact that they were rare, for this suggested that they were not caused by multiple scattering as was presumably the case for the β-particles. This was a crucial point, for one must remember that even the thin gold foil used in the experiment had a thickness of about 2,000 atom diameters. With only one α-particle in 10,000

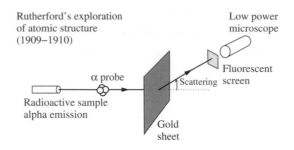

Fig. 1.1 Schematic representation of the Rutherford experiment. The α-particles consisted of helium nuclei composed of two protons and two neutrons. Although the gold foil constituting the target was as thin as possible it nevertheless had a thickness of about 2,000 times the diameter of an atom. Therefore the question of whether or not the α-particles experienced multiple scattering was a crucial issue. Large scattering angles had already been observed in earlier experiments, in particular with β-particles, but they were attributed to multiple scattering. So what made the Rutherford experiment remarkable was not the fact that large scattering angles were observed but rather the fact that the cross section of the α-particles was small enough to exclude multiple scattering.

being diffracted there was indeed a good chance that no multiple scattering had occurred. In short, instead of producing multiple scatterings which were difficult to interpret, this experiment produced much "cleaner" results. It is in this sense that the Rutherford experiment was a prefiguration of the accelerator experiments which were carried out in subsequent decades. The three elements of an accelerator experiment were already there, namely (i) a source emitting an energetic stream of particles, (ii) a target and (iii) a detector allowing measurement of the scattering angle.

What are the implications of the Rutherford experiment for the design of quasi-experiments in the social sciences? First of all, its success was ensured by selecting the "right" probe. Because α-particles were positively charged they were repelled by the gold nuclei and because they were sufficiently energetic they were able to probe the tiny nucleus.[2] This statement is easy to make with the wisdom of hindsight, but in 1909 it was not immediately obvious that the observation could not be interpreted in the framework of the prevailing plum pudding model. After all, even in this model the charge of the atom was concentrated in the plums and was not distributed uniformly. As a matter of fact, it was only two years after the experiment that Rutherford proposed a model in which the positive charge was concentrated in only one plum, the nucleus of the atom (Rutherford 1911). One should keep in mind that Geiger, Marsden and Rutherford carried out similar

---

[2] The Rutherford experiment did not yield any insight into the electron cloud which was of course known to exist to make the atom electrically neutral. For that purpose the energy of the probes was just too high.

measurements with other metallic targets: aluminum, copper, iron, lead, platinum. Moreover, after its results had been published the Rutherford experiment was repeated by several other groups. Not only were its observations confirmed but they were also tested in a broad range of experimental conditions. In spite of the novelty of the Rutherford experiment, one should not expect a real breakthrough to result from a single experiment; it was rather one link, albeit a crucial one, in a long chain of experiments. Cathode rays, X-rays, β- and α-particles all contributed to progressively shape and sharpen the picture. These efforts were productive and fructuous because they were focused on a single, well-defined question: what is the structure of the atom and what are the interactions between its different components? In contrast, in the social sciences, investigations are scattered over a broad range of questions. As a result the process of accumulation of knowledge and understanding works very poorly. It is one of the main objectives of this book to show that bonds and interactions between social agents play a key role and therefore deserve extensive investigation. This objective can provide a unified purpose and framework for various contributions and thus start a process in which knowledge can be produced in a cumulative way, as is the case in physics.

As far as the methodology of the Rutherford experiment is concerned it can be adapted to the social sciences without much difficulty, at least in its principle. Any shock, modification or mutation can be used as an experimental probe. However, it is much more difficult to define a set of *calibrated probes*, by which we mean a set of shocks whose intensity and duration can be controlled. In a subsequent chapter, the attack of September 11, 2001 will be considered as a probe and we examine how suicide rates responded to this event. It will be seen that September 11 had no measurable impact on suicide rates in the United States.[3] In short, big events such as September 11 do not provide good probes for the phenomenon of suicide. But big events can be used as probes for other phenomena. For instance, the destruction of the Ayodhya mosque in northern India in December 1992 brought about considerable disturbances between Hindu and Muslim communities in countries as diverse as Afghanistan, Britain, Canada and Pakistan.[4] This suggests that events of this kind are good probes for investigating the interaction between religious communities. However, even in this case we do not have a set of calibrated events, but rather we have a small number of events whose intensities can only be estimated on a fairly qualitative basis. We come back to this question in the next chapter.

---

[3] In a sense this absence of impact is similar to what is observed in the Rutherford experiment at the level of 99.990% of the particles which go through the target without being deflected. Unfortunately, in the case of suicides the accuracy of the measurement is too poor to identify a possible effect as small as the 0.01% deflected particles in the Rutherford experiment.

[4] For more details see Roehner (2004, Chapter 4) or Roehner, Sornette and Anderson (2004).

It must be emphasized that test-probes do not need to be shocks but can also take the form of structural changes. In the next section we give a number of examples of test-probes of this kind.

1.2 Boiling points as test-probes

In sociological language, evaporation and boiling can be seen as a drop-out phenomenon. When a molecule which is close to the surface of the liquid has enough kinetic energy (as a result of thermal agitation) it can overcome the attraction of surrounding molecules and escape from the liquid. When the process occurs at temperatures below the boiling point it is the phenomenon of evaporation. At boiling temperature the drop-out phenomenon spreads from the vicinity of the surface to the bulk of the liquid. Thus one expects the boiling temperature to have a strong connection with the strength of intermolecular interactions: the stronger the interaction, the higher the boiling point. This relationship is illustrated in Fig. 1.2a. It can be seen that for alkanes the boiling temperature increases along with molecular weight. For alkanes there is a direct relationship between molecular weight and intermolecular interaction (see Fig. 1.2b). However, this is not true

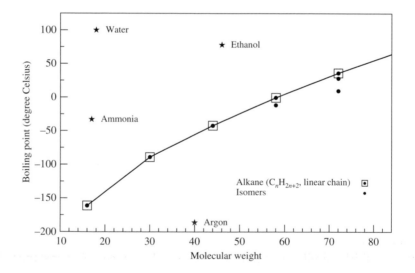

Fig. 1.2a Boiling point as a function of molecular weight. It is fairly easy to explain why the boiling point is connected with the strength of the molecular attraction, but it is far more difficult to explain this attraction in terms of the characteristics of the molecules. For alkanes, molecular attraction happens to be proportional to molecular weight (see Fig. 1.2b), but this rule holds only for alkanes. Even for their isomers (i.e. those alkanes which do not have a linear chain) the rule is only roughly true and it is not true at all for the substances indicated by stars. *Source: Lide (2001).*

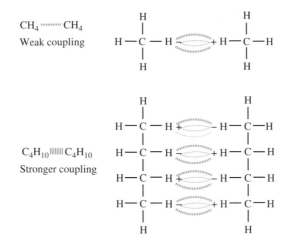

Fig. 1.2b Attraction between alkane molecules. The figure shows two alkanes: methane (CH_4) and butane (C_4H_{10}) The attraction due to induced dipoles (also called London bonds) represented in the figure is a form of interaction which is common to all molecules, but besides this mode most molecules have several other modes of interaction. The particularity and simplicity of the alkanes come from the fact that they do not have other modes of interaction. This is the reason why the attraction between two alkane molecules is proportional to their number of hydrogen atoms $2n + 2$ and thus to their molecular weight $14n + 2$.

for other compounds, as shown by the fact that argon and ethanol have very different boiling points in spite of having almost the same molecular weight. The low boiling temperature of argon is not surprising because we know that all rare gases have weak intermolecular interactions. On the other hand, the high boiling point of ethanol can be understood because we know that ethanol molecules are bound together by strong hydrogen bonds. Needless to say, this understanding relies on the fact that intermolecular interactions have been explored by various methods. It is because we know the properties of rare gases and of hydrogen bonds that we are able to interpret the outliers in Fig. 1.2a and to realize that boiling temperature indeed provides a reliable measure of intermolecular bonds. Back in the nineteenth century this realization was by no means obvious. Nowadays we are in a fairly similar situation in the social sciences. For instance, it has been presumed (in particular by Emile Durkheim) that suicide rates have a strong connection with the strength of social bonds, but unless we can measure this strength in an independent way such a claim is nothing more than a hypothesis or a tautology.

Societies are by no means homogeneous but comprise several different kinds of social agents. This observation suggests that it can be of interest to take inspiration from the physics of mixing.

1.3 Basic principles of the physics of mixing

The rationale of using the properties of mixtures as a way of exploring bonds can best be explained through an example taken from the ecology of animal species. Consider a population A of rabbits and a population B of squirrels. These populations to some extent share the same habitat and therefore they do interact; however, as they feed on different sources of food their interaction is for the most part of the non-competitive kind. Consequently, the death rate of rabbits will be fairly independent of the number of squirrels; in other words, the plot of the death rate $d_A(x)$ of the rabbits as a function of the percentage x of the squirrels in the combined population $A+B$ will be close to a horizontal line. Moreover, a plot of the total death rate $D(x) = d_A(x) + d_B(x)$ will also be a straight line, albeit not a horizontal line; it joins the two points $(0, d_A = d_A(0))$ and $(1, d_B = d_B(1))$, and its equation is $D = (1-x)d_A + xd_B = x(d_B - d_A) + d_A$.

Now, replace the squirrels by a population of foxes. This time there will be interactions between the two populations. As a result of these interactions the plot $d_A(x)$ of the death rate of the rabbits will no longer be a horizontal line, but rather an upgoing curve. Similarly, the plot of $D(x)$ will no longer be a straight line but rather a curve joining the points $(0, d_A)$ and $(1, d_B)$.

It is an easy step to shift from this example to physics. Consider a mixture of two liquids A and B. We replace the rabbits by the molecules A, the squirrels by the molecules B, the percentage x by the proportion of molecules B in the liquid, $x = n_B/(n_A + n_B)$, where n_A, n_B are the numbers of A and B molecules respectively, and the death rate of the rabbits by the rate at which the molecules A leave the liquid. This rate defines the partial pressure $p_A(x)$ of A in the vapor above the liquid. If the molecules A and B have no interaction, the plot of $p_A(x)$ will be a horizontal line. Moreover, the plot of the total pressure $P(x) = p_A(x) + p_B(x)$ will be a straight line joining the points $(0, p_A = p_A(0))$ to $(1, p_B = p_B(1))$. Its equation will be the same as previously: $P^{\text{ideal}}(x) = x(p_B - p_A) + p_A$. As examples of liquids which display such behavior one can mention methanol and ethanol, pentane and hexane and, in a more general way, pairs of liquids whose intermolecular interactions A–A and B–B are fairly similar.[5]

If one replaces the previous ideal liquids by pairs of liquids whose molecules interact in new ways, $p_A(x)$ will no longer be a horizontal line and $P(x)$ will no longer be a straight line; both lines will be replaced by curves. Naturally, the shapes of these curves depend upon the interaction; conversely, it must be possible

---

[5] This expresses what is called Raoult's law for ideal solutions. Note that in this context the word "ideal" is not used in the same sense as in the expression "an ideal gas". In the latter case it means that the molecules of the gas do *not* interact (i.e. have few collisions which last only short times). Here, in contrast, there *are* A–A, B–B and A–B interactions; in the absence of A–B interactions the two liquids would simply be unable to mix. Yet, as in the parallel with the rabbit–squirrel populations, these interactions are not innovative and disruptive to the point of radically changing the molecular organization of the mixture.

(but not necessarily easy) to infer information about the interaction from the shape of these curves. For instance $P(x) < P^{\text{ideal}}(x)$ signals a strong $A–B$ interaction whereas $P(x) > P^{\text{ideal}}(x)$ indicates a rather weak $A–B$ interaction. An example of the first kind is ethyl ethanoate, $C_4H_8O_2$, and trichloromethane, $CHCl_3$; these liquids are said to have a negative deviation from Raoult's law. An example of the second kind is cyclohexanone, $C_6H_{10}O$, and ethanol, C_2H_6O: these liquids are said to present a positive deviation from Raoult's law.

The technique of mixing two populations with the purpose of exploring their interactions will be applied to social phenomena in a subsequent chapter.

1.4 Immiscible liquids

There is an important case that we did not consider so far. Suppose that the molecules of liquid A (as well as those of liquid B) have strong interactions, but that the molecules A and B interact very weakly. What happens in this case when the two liquids are mixed? Observation shows that they will remain immiscible. Actually, this is a very common situation; familiar examples are: oil–water, hexane–water and acetone–glycerol. The reason for the immiscibility is easy to understand. When a molecule B is positioned between two molecules A this results in an increased distance and therefore in a substantially higher $A–A$ potential energy of interaction between A molecules (to use the image used previously for star formation, one can say that the spring which represents the interaction is more stretched). The same reasoning naturally applies when a molecule A is positioned between two molecules B. As the $A–B$ interaction is weak, these reductions in interaction (corresponding to increases in potential energy of interaction) will not be offset by the creation of additional $A–B$ interactions. In short, the mixed configuration would have a higher energy than the configuration in which the molecules A and B keep their initial, unmixed configuration.[6] Although this reasoning is fairly clear it remains rather qualitative. Actually, it is difficult to predict whether two liquids will be miscible or not.[7]

For ternary solutions the picture may be even more complicated. For instance, glycerol and acetone are almost immiscible, but if a sufficiently large quantity of water is added all three components become miscible. Water acts as a facilitator for the mixing of glycerol and acetone.

---

[6] In the previous parallel with ecological systems this would correspond for instance to populations of rabbits and trout. This, however, is a fairly trivial example in the sense that the two species do not interact because their habitats are completely different. A more relevant example would be two populations of ants (either of the same or of different species) which establish different non-interacting nests.

[7] It may be observed that immiscibility is never total. Even two liquids such as hexane and water, which are said to be "immiscible", are in fact miscible in very small proportions. Our previous argument does not account for this fact.

What are the analogues of these phenomena in the social sphere? One may speculate that the phenomenon of integration of a population of immigrants B into a society A is brought about by the creation of A–B links and that this is usually easier when A and B have the same kind of interactions. At the other end of the spectrum, the formation of ghettos can be seen as a case where the A–B interactions are too weak to overcome the stronger A–A and B–B interactions. However, so long as we are not able to measure social interactions in a reliable way, such arguments are nothing more than hypotheses and tautologies.[8]

In the next section the previous ideas are extended to a broader range of physical properties.

1.5 Physical properties of a mixture as a tool for exploring bonds

In the previous section we considered partial vapor pressures over a mixture of liquids. It is important to realize that any physical variable attached to a mixture can provide information about the interactions which are at work. The graphs in Fig. 1.3a, b, c give three examples for the case of a mixture of water and ethanol. Fig. 1.3a shows that the density of the mixture differs from a straight line, thus revealing an interaction between water and ethanol molecules. The fact that the curve is above the straight line (corresponding to the case of an ideal solution) shows that the mixing is accompanied by a contraction which reveals a strong A–B interaction. Depending on the variable that one considers, the non-linearity may be more or less apparent. For the velocity of sound (Fig. 1.3c) it is about six times larger than for the density. Can we, at least qualitatively, understand why the velocity of sound is higher in the mixture? This requires of course an understanding of the mechanism of sound propagation. For sound to propagate, successive layers must be put into motion. If there are strong intermolecular bonds there will be two (competing) effects:

- In a given layer strongly coupled molecules will behave as a massive cluster of molecules. This increases inertia effects and tends to reduce the velocity of sound.
- The strong bonds will facilitate the transmission of the wave from one layer to the next. This effect tends to increase the velocity of sound.

Which one of these effects will prevail is not easy to determine. Observation shows that the first effect prevails in gases while the second prevails in liquids.[9] In the case of a water and ethanol mixture the maximum of the velocity occurs for a concentration by weight of ethanol of 27%. This percentage is close to the

---

[8] Of course, there is a fundamental difference in the sense that, in contrast to molecules, people can adapt to new situations. However, adaptation often relies on the replacement of generations; it should not be overestimated for time spans shorter than 20 years.

[9] More details on this point can be found in Roehner (2005a).

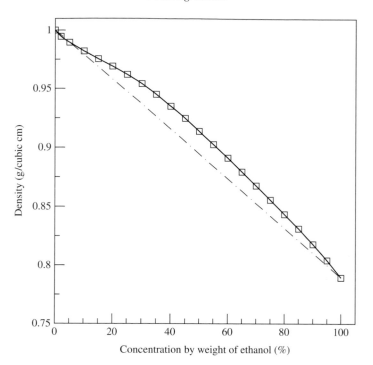

Fig. 1.3a Density of a mixture of water and ethanol. In the absence of any (new) interaction between the molecules of water and ethanol the density of the mixture would coincide with the (dashed) straight line. The non-linearity reveals the interaction, but in the case of the density the deviation remains fairly small (less than 2.5% at its highest point). *Source: Landholt-Börnstein, Group IV, Vol. 1, Part b: Dichten binärer wässeriger Systeme und Wärmekapazitäten flüssiger Systeme [Density of binary solutions and heat capacity of liquids]. Berlin: Springer, 1977; see also Holmes (1913).*

proportion for which the mixing produces the highest heat release, which can be considered as signaling a peak in the A–B interaction (Bose 1907). The density curve suggests that the inertia effect is probably small, thus it seems that the maximum in sound velocity can indeed be explained on the basis of the second effect.

This example shows how it is possible to extract information on interactions from experimental properties of mixtures. Nowadays, this approach is no longer used because more direct methods are available, such as small-angle X-ray or neutron scattering. However, prior to the introduction of these techniques, that is to say prior to 1925 approximately, reasoning based on the properties of mixtures was a standard tool for exploring intermolecular forces. Table 1.2 recapitulates various methods which have been (or still are) in use for this purpose.

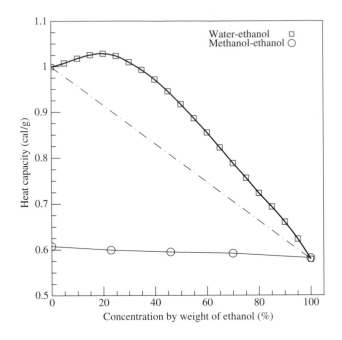

Fig. 1.3b Heat capacities of mixtures of liquids. There is a sharp contrast between a mixture of water and ethanol whose curve is highly non-linear and a mixture of methanol and ethanol whose curve is almost a straight horizontal line. In the first case, the mixture brings about new interactions, whereas in the second the A–B interaction is identical to the A–A and B–B interactions. *Source: Same as Fig. 1.3a.*

1.6 Estimating the correlation length in social phenomena

In physics the degree of collective behavior that exists in a system is often characterized by the correlation length ξ. This parameter can be measured from light scattering cross sections $f(q)$ through a fit of the so-called Ornstein–Zernike plot $f(q) = f_0/(1 + q^2\xi^2)$, where q is the wave vector. The correlation length does not directly provide an estimate of the interaction strength, it is rather a characterization of the degree of cohesion in the motion of the molecules. The degree of cohesion in turn is the result of two competing effects: (i) interactions tend to induce more cohesion; (ii) random thermal fluctuations tend to increase the standard deviation and to reduce the cohesion. If the second factor can be considered as constant, the correlation length will give us at least relative estimates of the interaction strength. It is therefore worthwhile to wonder how it is possible to measure correlation length for social systems. This is the question that we examine in the present section.

Correlation analysis is a standard tool in econometrics and in several other fields of the social sciences. Over the past decades it has been improved in many

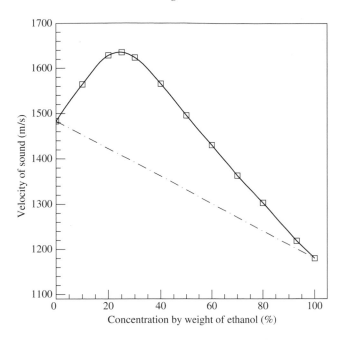

Fig. 1.3c Velocity of sound in a mixture of water (*A*) and ethanol (*B*). The strong non-linearity of the curve shows that the velocity of sound crucially depends upon the *A–B* interactions. Provided one has some understanding of the mechanism of sound propagation, it will be possible to use this curve as a probe of the *A–B* interactions. *Source: Nozdrev (1965, pp. 354–7).*

respects: (i) introduction of correlations with arbitrary time lags; (ii) cointegration techniques for analyzing time series with a common trend; (iii) introduction of techniques for analyzing time series with time varying standard deviations. However, none of these developments addressed the question of how correlation analysis can be used to describe the collective motion in a system of social agents. This question involves in fact two different steps. The first requirement is to select a set of time series for which one has good reasons to think that the underlying phenomena share the *same* interaction. The second question is to define a statistical indicator which can describe the degree of collective motion.

Possible ways to satisfy the first requirement will be illustrated through a few examples at the end of this section. A possible answer to the second question is given in Fig. 1.4. The procedure is simple and straightforward. Suppose that 1, 2, 3 and 4 represent four spatially separated markets on which transactions in a given commodity are conducted. The degree of cohesion of these markets (often referred to as the degree of market integration) is affected by two phenomena. (i) Price arbitrage tends to increase the cohesion. For instance, if at a given moment the price p_2 is higher than the price p_1 there will be a transfer of goods from

Table 1.2 *Experimental means for the exploration of bonds*

Field	Measurement	Key mechanism	Absolute (A) or relative (R) measurement of bond strength
Physics			
1	Boiling point	Rate at which molecules leave the liquid	R
2	Miscibility or immiscibility of two liquids *A* and *B*	Is interaction *A* compatible with interaction *B*?	R
3	Light scattering	Fourier transform of two-point correlation functions	R
4	Autocorrelation of the velocity	Exploration of transport phenomena	R
5	Small-angle neutron scattering	Exploration of alloy segregation	A
6	Infrared spectroscopy	Exploration of atomic interactions within molecules	A
Social sciences			
1	Mixing of two populations *A* and *B*	Is interaction *A* compatible with interaction *B*?	R
2	Correlation length	Exploration of the interactions between commodity markets	R
3	Normalized correlation length	Exploration of interactions between share prices	R
4	Autocorrelation of prices	How fast is equilibrium restored?	R
5	Severance of social ties	Increase in suicide rates	R

Notes: Any physical phenomenon which depends upon the strength of intermolecular forces can be used as an experimental probe for exploring molecular bonds. Thus, while the compression of dilute gases does not give any indication about the (small) inter-molecular interaction, the compression of dense gases provides information about the attraction of molecules through the van der Waals equation of state. Examples illustrating the use of some of these methods in the social sciences will be given in subsequent chapters. It can be noted that the methods based on pair correlations can be used only for macro-systems such as countries, regions or markets for which time series are recorded. As there are no recorded time series at the micro-social level of individual agents, one cannot estimate their pair correlations in this way. As a matter of fact, the situation is fairly similar in physics, in the sense that the motion of individual molecules cannot be measured; nevertheless, the correlation function of pairs of molecules can be estimated because one is able to establish a connection with the intensity of light scattering.

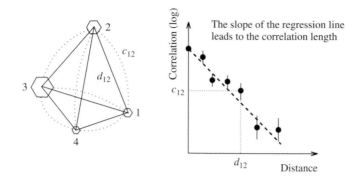

Fig. 1.4 Definition of the correlation length of an economic system. The hexagons schematically represent four commodity markets; d_{12} is the distance between markets 1 and 2; c_{12} represents the pair correlation between the price series $p_1(t)$ and $p_2(t)$ observed on markets 1 and 2 respectively. Observation reveals that there is a marked correlation between the pair correlation c_{kl} and the distances d_{kl}. The fact that the plot turns out to be fairly linear shows that the relationship can be represented by the formula $c_{kl} = \exp(-d_{kl}/\xi)$ where the parameter ξ is the correlation length.

1 to 2, thereby reducing the price gap. (ii) Local shocks which generate random fluctuations tend to decrease the degree of cohesion.[10]

The measure of the correlation length is based on the correlations between price series. For each of the six pairs of time series (p_k, p_l) one computes the cross correlation ρ_{kl}; a plot can then be drawn which represents the ρ_{kl} as functions of the respective distances between markets. The correlation length ξ can be defined in the same way as in statistical mechanics by the formula

$$\rho_{kl} = \exp(-d_{kl}/\xi) \tag{1.1}$$

Consequently, if the slope of the regression line of the $(d, \ln \rho)$ plot is denoted by $-m$, the correlation length will be given by $\xi = 1/m$. If distances are expressed in kilometers, the correlation length will also be expressed in the same unit. This provides an absolute measure through which it is possible to compare the degrees of integration of different sets of markets.

Estimates of the correlation length in a number of cases are shown in Table 1.3. It appears that their relative magnitudes are in agreement with what would be expected intuitively. For instance, the correlation length is larger for ports than for inland markets, which would be expected on account of the fact that transportation by sea was (and still is) less expensive than transportation by land.

---

[10] The system is also subject to *global* exogenous shocks, for instance due to changes in business conditions or to climatic changes. Such global shocks tend to increase the *apparent* cohesiveness of the system. They are omitted in the present discussion for the sake of simplicity.

Table 1.3 *Correlation length of commodity markets*

Market	Year	Number of markets	Correlation length (kilometer)	Correlation between distances and correlations
Wheat markets				
1 Bavaria	1829	9	31 ± 5	0.85
2 France (center)	1873	11	55 ± 18	0.87
3 France (ports)	1873	10	211 ± 41	0.58
4 United States	1970	31	560 ± 40	0.70
Potato markets				
5 Prussia	1837	10	12 ± 4	0.63
6 United States	1970	31	96 ± 27	0.50

Notes: The nineteenth-century data are for wheat markets established in different places (at this time there were wheat markets in all cities or towns of some importance); the US prices are average sale prices in different states. The year indicates the center of the time window. The last column gives a measure of the goodness of fit in terms of the correlation of the scatter plot represented schematically in Fig. 1.4b. The fact that the correlation length is smaller for potatoes than for wheat makes sense in so far as potatoes have a lower price per unit of mass, which makes their relative transport cost higher. Cases 1, 2, 3 correspond to monthly price series whereas 4, 5, 6 correspond to annual price series. More details can be found in Roehner (1995, Chapter 6; 2000a, pp.181–4; 2000b).
Sources: 1: Seuffert (1857); 2, 3: Drame *et al.* (1991); 4: Langley and Langley (1989); 5: Engel (1861); 6: Lucier *et al.* (1991).

Clearly the expression (1.1) requires that $\rho_{kl} > 0$. In social phenomena it is not rare to observe negative correlations (in particular because the level of noise is higher). In those cases, it is sensible to replace the exponential by its expansion to first order and to define the correlation length ξ as

$$\rho_{kl} = 1 - d_{kl}/\xi \qquad (1.2)$$

Remark   If in addition we normalize the distances so that the largest distance d_m becomes equal to 1, Equation (1.2) turns into

$$\rho(x) = 1 - x/\xi' \qquad \xi' = \xi/d_m \qquad x = d_{kl}/d_m \qquad 0 \le x \le 1 \qquad (1.3)$$

So far we have considered spatially separated markets. This condition represents a serious limitation in so far as it excludes financial markets.[11] A natural question, therefore, is whether it is possible to define a correlation length for

---

[11] Even when financial markets are spatially separated, geographical distance plays no role because there are no transport costs.

non-spatial time series. In fact, the previous procedure need only be changed slightly. The modification which has to be introduced relies on the observation that large distances are associated with small correlations (see Fig. 1.4). Therefore, if we sort the correlations in decreasing order and plot them as a function of their rank order, the resulting graph will be very similar to the one in Fig. 1.4. For instance, the point (d_{12}, c_{12}) will be replaced by the point $(4, c_{12})$ where 4 is the rank order of c_{12} (excluding $\rho = 1$, which is the correlation of each market with itself). To make the graph independent of the number of pairs, the point $(4, c_{12})$ can be replaced by the point $(4/6, c_{12})$. Observation shows (see Table 1.4a) that this procedure leads to an alternative correlation "length" (to be denoted by δ to distinguish it from ξ) which is almost identical to ξ up to a multiplicative factor. In short, δ will be defined by the following relation:

$$c_{kl} = 1 - q/\delta_1 \qquad q = i/p \qquad (1.4)$$

where i denotes the rank order of c_{kl} in the decreasing sequence of correlations and p denotes the total number of pairs, $p = n(n-1)/2$ where n is the number of markets.

The correlation length can also be defined in another, fairly equivalent, way which rests on the average cross correlation:

$$\rho_m = \frac{1}{p} \sum_{k \neq l} \rho_{kl}$$

Integrating (1.3) with respect to x, one gets

$$\rho_m \simeq \int_0^1 \rho(x)\mathrm{d}x = \int_0^1 \left(1 - \frac{x}{\xi'}\right)\mathrm{d}x = 1 - \frac{1}{2\xi'}$$

Replacing ξ' by the notation δ_2 (in line with the notation δ_1 introduced above) one gets the definition

$$\delta_2 = \frac{1}{2(1 - \rho_m)} \qquad (1.5)$$

Table 1.4a shows that the expressions (1.4) and (1.5) lead to fairly similar results.

Illustrative examples Table 1.4b shows the variations in correlation length in the upgoing and downgoing parts of a price peak. For wheat markets (and in a more general way for commodity markets) the maximum of the correlation length occurs in synchronicity with the maximum of the price. In contrast for stocks

Table 1.4a *Normalized correlation length computed in four different ways*

	Regression between dist. – log(cor) (1.2)	Regression between dist. – cor (1.3)	Regression between rank order – cor (1.4)	Mean of cross-correlations (1.5)
Bavaria (wheat markets)	19.7	20.3	21.1	20.3

Notes: The numbers in parentheses refer to the equation numbers in which the correlation length is defined. The calculation was performed for monthly wheat prices on 13 Bavarian markets over the time interval 1815–55. The term "normalized" used in the title of the table refers to the fact that the x variable is restricted to the interval $(0,1)$. Whereas the definitions (1.2) and (1.3) apply only to spatially separated markets, the two other definitions can be used for any set of time series which describe an intensive variable such as stock prices, marriage rates, unemployment rates, etc. The good agreement between definitions (1.2) and (1.3) on the one hand and definitions (1.4) and (1.5) on the other hand shows that the latter can be considered as extensions of the standard definition.
Source: Bavarian wheat prices: Seuffert (1857).

Table 1.4b *Illustrative application of the normalized correlation length: variation in market cohesion during a price peak*

French wheat markets	1844	1847	1850
δ_2	8.5	300	9.2
Dow Jones stocks	1926	1931	1934
δ_2	2.1	12	2.1

Notes: The two time intervals were marked by price peaks. During the first episode prices peaked in late 1847, during the second they peaked in mid 1929. For wheat prices the turning point occurred simultaneously on all markets (or more precisely in the same month). This resulted in a considerable increase of the normalized correlation length. In contrast, for stock prices the real turning point differed from share to share; the highest market synchronicity was reached during the downward phase as can be seen from the fact that the normalized correlation length is maximum in 1931 rather than in 1929 or 1930. The wheat prices are fortnightly data for 11 markets, the share prices are monthly data for 9 stocks from the Dow Jones average.
Sources: Drame *et al.* (1991), Roehner (2001, Chapter 10).

the turning points do not occur at the same moment; in 1929 and 1930 δ_2 is on average equal to about 5 and rises to 12 only in 1931.[12]

---

[12] Note that it would not make sense to compare the magnitude of δ_2 on different markets. For wheat markets, for instance, δ_2 depends upon the maximum distance d_m; the larger d_m, the lower the correlations for the most distant markets; thus if the sample of markets is extended δ_2 will decrease so to say mechanically. In other words, δ_2 is sample dependent and the only comparisons that make sense are those which are restricted to the same sample. This is of course the case for the variations in the course of time considered in Table 1.4b.

We close this chapter by giving a brief outline of the different kinds of bonds that we study in subsequent chapters.

1.7 Outline of the book

Figure 1.5 gives a schematic representation of the hierarchy of bonds that can be found between and within different societies.

In Part II we study influence and control phenomena which occur at the macro-level of states and societies. For instance we will try to identify the factors which determine the Zeitgeist (that is to say the prevailing paradigm) of a society. We also try to understand through what means and channels a state can influence another. We explore the consequences of a weak interaction between management and execution levels. In this part, our exploration will be mostly of a qualitative nature. Our main objective is to identify mechanisms which have so far been relatively overlooked and neglected. This study paves the way for more quantitative investigations.

In Part III we study bonds at the micro-level of family units. Most of our exploration concerns the phenomenon of suicide but instead of studying suicide for itself we reverse the perspective and use suicide rates as a means for gauging the

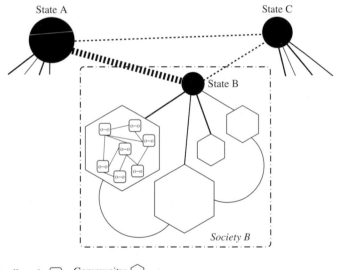

Family unit: [o–o]   Community: ◯

Fig. 1.5 Schematic representation of bonds and interactions at macro- and micro-level. Ties and interactions which concern states and societies globally are referred to as macro-bonds whereas ties at the level of family or community units are referred to as micro-bonds. Part II of this book is mainly concerned with macro-bonds, whereas Part III is devoted to micro-bonds.

strength of bonds. It will be seen that every time specific bonds are severed there is an ensuing surge in suicide rates. Conversely, rising suicide rates signal declining or eroding social links. The starting point of our investigations is the observation that suicide rates are higher for unmarried people than for married people. We systematically explore the various testable implications of this assertion: (i) in groups whose sex ratio notably differs from one; (ii) in populations in which there has been a rapid change in the average age of marriage; (iii) for individuals who are in a situation of social isolation, such as inmates or persons belonging to minority groups. Finally, we investigate the phenomenon of apoptosis, which can be considered as a parallel of suicide in the biological world.

Appendix   Gauging the links between two twinned communities

This appendix describes the links established between the populations of two small towns, one located in France and the other in southern Germany. The objective of this case study, which is based on personal observation, is to show how it is possible to go beyond mere appearance in evaluating cross-national bonds.

The practice of what is called "town twinning" in Europe or "sister cities" in the United States refers to the pairing of two towns or cities from different countries with the purpose of fostering human contacts, closer understanding and promoting cross-border projects. Each twinning provides a testing ground for how closer links can be established between two populations and how it is possible to estimate the strength and density of the network of links which have been established.

Twinning is of little significance for big cities; for instance, Paris is twinned with Rome, something probably less than one percent of Parisians know about. However, one would expect twinning to be of more significance for the residents of small towns. As a case in point we consider the following towns:

- Beaulieu-sur-Dordogne in the center of France (200 kilometers to the north of Toulouse) which had 1,500 residents in 2000.
- Scheinfeld in south-east Germany (200 kilometers to the north of Munich) which had 3,000 residents in 2000.

The towns were twinned in 1981 and in 2006 they celebrated the 25th anniversary of their pairing. On that occasion about 150 residents from Scheinfeld went to Beaulieu during the Easter weekend. An equal number of people from Beaulieu were present at the various gatherings which took place during these three days. These 300 people represented 6.7% of the combined population. Transposed to the case of Paris and Rome, such a proportion would mean exchanges involving more than one million people and would represent a major event. Moreover, the fact that the towns had been twinned for 25 years suggests that enduring ties may have been set up in the course of time. Are these expectations confirmed by observation?

The main obstacles to the establishment of close ties are the language barrier and the different cultural environments. The following observations indicate to what extent these obstacles were surmounted.

- Music affords a means of communication between people who do not speak the same language and this was well understood by the organizers. Scheinfeld came with two orchestras and Beaulieu staged a choir and a town band. However, the repertoires of the orchestras had no overlap; Scheinfeld's "Stadt Kapelle" and Beaulieu's band did not play a single piece together, not even the European anthem.
- Each town had a few perfectly bilingual persons who were able to translate the welcome speeches. Nevertheless, the ability of the participants to communicate remained limited. Very few mixed groups formed for the visit to the town and the surroundings.
- At least half of the participants on both sides were institutional participants: organizers, members of the municipal councils, musicians in the orchestras. No more than 50 common citizens seem to have taken part, that is to say about 1% of the total population. The fact that only three or four houses were decorated with flags of the two countries confirms the low level of participation.
- Cross-national communication seems to have been confined to three areas: religious celebration, soccer and food tasting sessions. The fact that Scheinfeld belongs to the Catholic part of Germany together with the fact that the meeting took place during the Easter weekend explains why the mass on Sunday morning was officially mentioned on the program. The European soccer championships have become a center of interest for the media and for the public, therefore it is natural that soccer offered a common "cultural" ground.

This case study suggests a number of criteria for judging the degree of inter-action. Many of them, e.g. the number of non-institutional participants, or the proportion of mixed groups during the visit to the town, could be expressed in a quantitative way without much difficulty. Naturally, to be of real interest these observations should be repeated for a substantial number of twinned towns or cities.

2

The battle against noise in physics

In this chapter and in the next we discuss what we believe is the main difference between physics and the social sciences, namely the high level of noise which spoils and often altogether forbids the measurement of many social variables. This might seem a fairly unconventional view. Usually many other differences are underscored, for instance the role of individual freedom, the diversity and complexity of human behavior and so on. More specifically it is often stated that (i) social systems are inherently more complicated in the sense that societies comprise many different kinds of agents and interactions; (ii) physical phenomena can be observed in the laboratory through experiments which can be repeated whereas social observations, it is argued, cannot be repeated. We do not mean that these differences do not exist but we believe that they are not as fundamental as may be the impression at first sight.

The first point can be readily dealt with by noting that the level of complexity depends on how much detail one wants to take into account. Seen at the molecular level, the dissolution of a piece of sugar in a cup of tea is a phenomenon of horrendous complexity in the sense that it involves many kinds of molecules and a great diversity of interactions. In contrast, for many social phenomena the precise nature of the agents is unimportant. For instance, the field of demography focuses on the numbers of birth and deaths, a perspective in which most other social features become irrelevant.

The second argument may at first seem to be rock solid, but it must be considered in the light of the two following remarks.

- Strictly speaking, an experiment cannot be repeated even in physics. Consider for instance the swing of a pendulum, which is one of the simplest experiments one can think of. If we measure its period of oscillation with a precision of 0.1 second, two successive experiments may indeed give identical results, say 1.2 second. However, if we are able to make high precision measurements with an accuracy of 10^{-6} second, successive measurements will lead to different results, say 1.200123, 1.200759 and 1.200023. There may be several reasons explaining these fluctuations. One can for instance invoke drafts, or vibrations of the ceiling of the building to which the pendulum is attached. These vibrations are due to many factors: people walking at the same floor or at the floor above, vibrations due to trucks, trains or even planes in the vicinity of the building. These are high frequency shocks, but there are also low frequency perturbations. Suppose for instance that the experiment is repeated a few hours later; due to the rotation of the Earth on its axis the positions of the Sun and Moon with respect to the pendulum will no longer be the same; this will modify the directions of their gravitational forces which will slightly change the period.[1] In short, the question of whether or not it is possible to reproduce an experiment really depends on the level of accuracy that one wishes to achieve.

- To complete our argument, it must now be shown that in the social sciences it is indeed possible to repeat quasi-experiments in a way which is similar to what is done in physics. For definiteness and in order to make the discussion more concrete we consider a specific phenomenon, the so-called Werther effect. According to this effect some people are induced to commit suicide when they learn of other people having committed suicide. This effect was first studied by David Phillips some thirty years ago in an influential paper (Phillips 1974).[2] More specifically, Phillips analyzed the fluctuations in monthly numbers of suicides in the United States in the month following the announcement of a suicide on the first page of the *New York Times*. For the time being let us accept Phillips's result that on average there is a 2.5 percent increase in suicides following the publication of suicide stories as compared to months in which no suicide story was published. In a system theory perspective there is a close parallel between the Werther effect and the swing of a pendulum. This parallel is schematically illustrated in Fig. 2.1. Obviously, the Werther experiment can be repeated as often as one wishes. It can be repeated in any year between 1851 (when the *New York Times* began to be published) and now. It can also be repeated in any country in which there is a newspaper similar to the *New York Times*. *The Times* of London or many other major newspapers published in European countries can be used in the same way, which means that the experiment can be repeated in several different countries.

There is however a major difference between the swing of a pendulum and the Werther effect. For the former the signal to noise ratio is much larger than one (ratios exceeding one hundred are not difficult to obtain), whereas for the Werther effect it is smaller than one.

---

[1] In technical terms this is called a tidal effect.

[2] It should be noted that in spite of numerous (but non-converging and non-cumulative) studies, the reality of this effect has not yet been established with certainty. We will come back to this point in the next chapter. Here we use this effect as an example of a quasi-experiment and the reality of the effect is irrelevant for the present discussion.

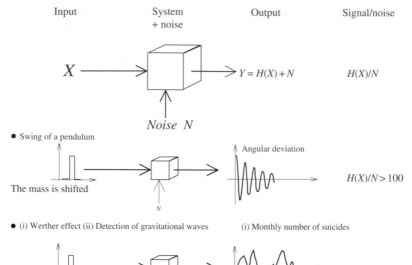

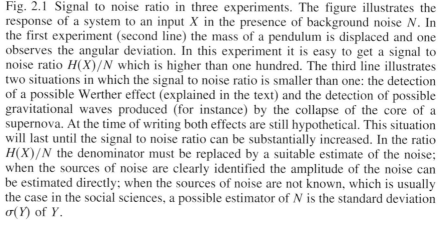

Fig. 2.1 Signal to noise ratio in three experiments. The figure illustrates the response of a system to an input X in the presence of background noise N. In the first experiment (second line) the mass of a pendulum is displaced and one observes the angular deviation. In this experiment it is easy to get a signal to noise ratio $H(X)/N$ which is higher than one hundred. The third line illustrates two situations in which the signal to noise ratio is smaller than one: the detection of a possible Werther effect (explained in the text) and the detection of possible gravitational waves produced (for instance) by the collapse of the core of a supernova. At the time of writing both effects are still hypothetical. This situation will last until the signal to noise ratio can be substantially increased. In the ratio $H(X)/N$ the denominator must be replaced by a suitable estimate of the noise; when the sources of noise are clearly identified the amplitude of the noise can be estimated directly; when the sources of noise are not known, which is usually the case in the social sciences, a possible estimator of N is the standard deviation $\sigma(Y)$ of Y.

   Throughout the history of physics and chemistry, signal enhancement and noise reduction have been (and still are) permanent concerns. It is natural therefore to draw on this knowledge and this is why we first consider two examples of physical phenomena. The first one is the familiar pendulum that we selected for its simplicity. In spite of this simplicity, it is a source of useful principles and guidelines. The second example, the detection of gravitational waves, is of interest because it is still awaiting a satisfactory solution. For this reason it provides a useful parallel with social science problems.

2.1 Improving the signal to noise ratio in the pendulum experiment

The signal to noise ratio Y/N can be raised either by increasing the signal Y or by reducing the noise N.[3] This might seem a fairly trivial remark, but it has far reaching consequences, especially for social phenomena. In the case of the pendulum, the step of increasing the signal is done almost without thinking about it. Nobody would give the pendulum an initial deviation of only two or three degrees. It is clear that with such a small amplitude the movement would be much influenced by friction at the articulation point or by air turbulence. On the contrary, if the pendulum oscillates with an amplitude of say 40 degrees, its energy is large with respect to the energy of perturbing factors, its movement will be more regular and the measurement of its period will be more accurate. This method of raising the signal to noise ratio plays an important role in different chapters of this book; for the sake of brevity we call it the *extreme value technique*. Naturally, it has inherent limitations. In the case of the pendulum the initial deviation cannot be made larger than 180 degrees,[4] or if one wishes to study the variations of the period as a function of the initial angle, small oscillations cannot be avoided. This raises the question of whether there are noise reduction techniques. From our previous discussion of the pendulum experiment we know that there are high frequency as well as low frequency sources of noise. How is it possible to reduce their impact on the movements of the pendulum? In the following we examine several kinds of means which can serve this purpose.

- If the experiment is performed in a vacuum container the mass of the pendulum will no longer be subject to high frequency fluctuations due to air turbulence.
- If in addition the experiment is performed overnight when the building and its surroundings are much quieter than during daytime, the fluctuations due to exogenous vibrations will be greatly reduced.
- The exogenous influence of the Sun and Moon cannot be eliminated; however, if the experiment is performed at times when the Sun and Moon are in the same configuration, this effect can be reduced.[5] Alternatively, since we know the strength of the gravitational forces and the way in which they influence the movement of the pendulum it is possible to compute appropriate corrections. Naturally, these corrections can only be made because we know the laws which govern this system. In the time of Galileo it would have been impossible to perform such calculations.

---

[3] Strictly speaking, the signal to noise ratio is rather (in the notations of Fig. 2.1) $H(X)/\sigma(N)$. However, when the noise N is of the same magnitude as the signal $H(X)$ it is difficult to know if there really is a signal; this is why we often replace $H(X)/\sigma(N)$ by $Y/\sigma(N)$. Such a substitution is particularly welcome for the Werther effect or for the detection of gravitational waves. For a system subject to high levels of noise $H(X)/\sigma(N) \ll Y/\sigma(N) \sim 1$; for a system subject to a low level of noise, $Y/\sigma(N) > H(X)/\sigma(N) \gg 1$.

[4] Initial deviations of more than 90 degrees require of course a pendulum with a rigid arm.

[5] The planets, especially Jupiter, may also have an influence but it is much smaller.

All these improvements will raise the signal to noise ratio and reduce the error bars.[6] Naturally, such high accuracy experiments have little in common with standard classroom experiments. What lessons can be drawn from this simple example? First, it must be stressed that the noise has been cut off at its source. This is an important point. In economics and in the social sciences the standard methodology is to select the data without giving much attention to the issue of noise reduction, and then to subject these noisy data to statistical analysis in the hope that a signal will emerge.[7] Unfortunately, no statistical treatment can substantially improve the signal to noise ratio, especially if the characteristics of the signal are unknown; it is only by enlarging the data set and by making use of additional information pertaining to the system under consideration that the signal to noise ratio can be raised markedly. The pattern matching technique that we examine below is an example of this kind. More generally, the pendulum example makes us realize that the battle against noise can only be won if we have good knowledge about the system under investigation and the sources of noise. Because they were unaware of the turbulence produced in the wake of the pendulum by the drag due to air resistance, physicists of the seventeenth century would have been unable to identify this source of noise. Similarly, building a device that can shield the pendulum from the vibrations of the building requires good knowledge of the physics of vibrations. Incidentally, this example illustrates the fact that all subfields of physics are connected: to make an accurate measurement of a phenomenon in the field of mechanics one needs a good understanding of fluid mechanics, astronomy or electroacoustics. The fact that in the social sciences many subfields are patently underdeveloped helps to explain why it is so difficult to perform accurate measurements.

The main point is that in physics, whatever their cost, the required improvements *can* be made. This leads us to the crucial question: how can one increase the accuracy of observations in the social sciences? Before coming to this point, we wish to discuss an example of physical measurement in which the problem of noise reduction is more serious than in the case of the swing of a pendulum.

---

[6] In the previous enumeration we mentioned only hardware improvements in the experiment itself. Naturally, for any experimental device the precision can be increased by making a large number of measurements and taking the average; the averaging technique will be considered in more detail later on.

[7] This assertion can be illustrated by the example of Phillips's paper about the Werther effect. Among the 33 suicide stories that he selected are cases as different as American actresses (Carole Landis and Marilyn Monroe) and various political figures including a Soviet defector (Victor Kravchenko), an Egyptian Field Marshal (Abdel-Hakim Amer) and a Chinese Army leader (Lo Jui Ching). All these events are considered indiscriminately which results in a high level of noise. Naturally, it is only through a better understanding of the phenomenon that a selection can be made which would raise the signal to noise ratio. The fact that an improved selection is possible is suggested by the observation that only a few cases contribute significantly to the signal.

2.2 Noise reduction in the detection of gravitational waves

Gravitational waves are ripples in the fabric of space-time. When they pass through a detector they will decrease the distance between two test-masses. This effect is very small however: 10^{-18} meter over a distance of 4,000 meters between the test-masses. In order to detect the effect of a gravitational wave the distance between the test-masses must be measured with a precision of 2.5×10^{-20} percent. In spite of being shielded in a concrete cover, the vacuum pipe which connects the test-masses is subject to many sources of vibrations: trucks on nearby roads, micro-earthquakes, sound waves due to supersonic planes, thermal variations and so forth.

In order to achieve this daunting objective two methods are used, which it is worthwhile to detail because they constitute two major techniques of signal detection. The first technique relies on pattern identification. It is a common experience that even in a very noisy place, such as a pub, we are able to identify a familiar melody however softly played. We do this by pattern matching. Our ears and brain are wonderful pattern matching organs. Catching gravitational wave signals is not unlike what our ears do routinely. Of course, the pattern of the signal depends upon the source. A supernova and a spinning neutron star have very different signal patterns. Moreover, for binary stars the period is specific to each pair of stars. As there are many other mechanisms which can trigger gravitational waves, it is easy to realize that the number of patterns to search can become very large, making pattern matching a formidable computational task. The second technique relies on using several detectors. As we already noted, micro-earthquakes or fluctuations in the measurement devices can cause disturbances that simulate gravitational events. Such factors are site dependent and are unlikely to happen at two widely separated sites. This is why there are five different sites: one in Germany, one in Italy, one in Japan and two in the United States.

Several lessons can be drawn from this example.

- It emphasizes the importance of having *different realizations* of the same phenomenon. If the sources of noise in these realizations are not positively correlated they will to some extent cancel one another, thus improving signal identification. An even more favorable case is when the sources of noise are negatively correlated, for in this case noise reduction can be much greater. This case, which unfortunately is not very common, is discussed in the next chapter.
- It shows how crucial it is to have some information about the expected shape of the signal. In the case of gravitational waves we have seen that there are many possible candidates, which means that a broad range of pattern matching tests must be performed. In the social sciences the situation is even more difficult in the sense that usually one knows very little about the shape of the signal that is to be expected. For instance, in the case of the Werther effect we do not know the timing of possible

excess-suicides. Actually, even if somehow we were able to know that the time scale of the phenomenon is a matter of days, it would be difficult to take advantage of this knowledge because in most countries only monthly suicide data are available.

- For the detection of gravitational waves it is of critical importance that the signal is as big as possible. Only phenomena of cataclysmic proportions have a chance of being detected. In this case we are in the same situation as in the social sciences in the sense that (i) the signal cannot be changed at will, and (ii) big events are rarer than small events. Thus, the detection of the phenomenon is conditioned by favorable circumstances which are out of the control of the researcher.

2.3 Pattern matching: a simulation

How can pattern matching be used in the social sciences? Figure 2.2 shows how the response of a system to an impulse looks when the level of noise is progressively increased. This simulation was performed by using the following linear stochastic equation:

$$Y_t = aY_{t-1} + N_t + \delta_{t,t_1} \qquad a < 1 \qquad (2.1)$$

where N_t is a Gaussian random variable of mean zero and standard deviation σ; the Kronecker symbol δ_{t,t_1}, which is non-zero only for $t = t_1$, describes the input shock. The first graph in Fig. 2.2 corresponds to $\sigma = 0$ and shows the purely deterministic response, a steep increase followed by an exponential fall. When σ is equal to 8 we see that the exponential fall is very distorted and no longer recognizable. However, it can be noted that the steep increase is still visible. Let us now forget for a moment that the graphs in Fig. 2.2 were generated by a simulation and assume instead that they correspond to a social phenomenon that one wishes to identify. The identification is highly conditioned by what we know about the shock. Parameters which are of particular importance are the time when the shock occurred and the sign of the shock. If these two parameters are known the shock can still roughly be identified on the three graphs in the second line of Fig. 2.2. This example clearly shows how important it is to have as much information as possible about the shock.

How does the simulation of Fig. 2.2 compare with the situation that one faces in the case of the Werther effect? Figure 2.3 shows the monthly fluctuations of suicide rates in the United States. The short spiky fluctuations are seasonal variations; it is well known that in the northern hemisphere suicide rates go through a maximum in May–June and through a minimum in December.[8] Because

---

[8] The seasonal pattern of suicides has been studied by Morselli (1879, pp. 136–42) and by Durkheim (1897, Chapter 3, Section 4).

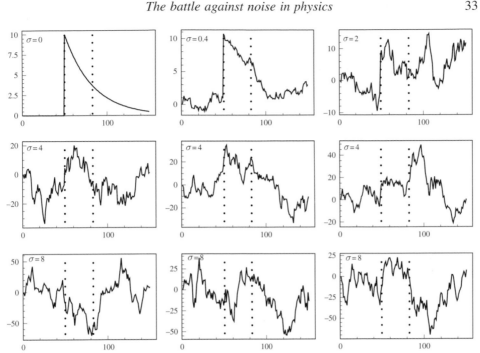

Fig. 2.2 Response of a first order system to a shock in the presence of increasing levels of noise. The figure represents the output Y_t of a system defined by the following stochastic recurrence equation:

$$Y_t = aY_{t-1} + \sigma\epsilon_t + \delta_{t,t_1} \qquad a = 0.97$$

where ϵ_t is Gaussian noise of mean zero and standard deviation 1 and δ_{t,t_1} represents a shock occurring at time $t_1 = 30$; σ takes the values 0, 0.4, 2, 4, 8. For $\sigma = 4$ and 8 the figure shows different realizations of the stochastic process in order to give an idea of its degree of variability. As σ is increased it becomes more and more difficult to identify the signal. In fact, signal identification crucially depends on how much information we have about the signal. If one knows that a sharp vertical discontinuity is to be expected it can be recognized even in the graphs with $\sigma = 8$. In contrast, the exponential relaxation process is much more difficult to identify in the last graphs. The vertical dotted lines indicate the relaxation time τ defined by $Y_t = e^{-t/\tau}$. Note that this relaxation time can also be observed very roughly in the overall shape of the broad peaks; it can be seen that their duration is approximately of the order of τ. This is of course not surprising because the response of the system to the shocks generated by the noise is of the same nature as its response to the deterministic shock δ_{t,t_1}.

of these huge seasonal fluctuations, the identification of the Werther effect is very problematic even if one knows the date of the suicide story published in the *New York Times*.

The techniques which have emerged from our discussion of physical experiments, namely the extreme value technique, the multi-observation approach and

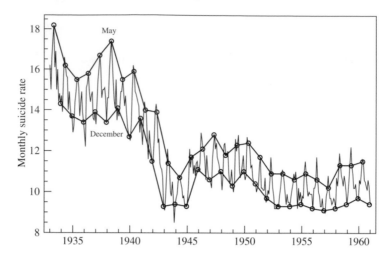

Fig. 2.3 Monthly variations of the suicide rate in the United States. The vertical scale refers to annualized monthly rates per 100,000 population. Monthly fluctuations (thin line) are seen to exhibit a seasonal pattern with a maximum usually occurring in May (represented by the upper curve) and a minimum in December (lower curve). The difference between seasonal maxima and minima was 27% in the 1930s but fell to 9% in the 1950s. Superimposed on these seasonal changes are substantial annual changes both in the seasonal pattern and in the average annual suicide rate. These changing features make the identification of any signal very difficult. *Sources: Linder and Grove (1947), Grove and Hetzel (1968).*

the various forms of pattern matching, will be examined in more detail in the next chapter in a social science context. They will be illustrated by several examples through which we will get a better feeling of their potential and limitations.

3

The battle against noise in the social sciences

Whenever the signal to noise ratio is smaller than one, identification is likely to be unclear, unconvincing and open to discussion. For instance, it is because the signal in the Werther effect is of the same magnitude as the noise that the very existence of this effect is still a matter of debate thirty years after Phillips's pioneering paper. Similarly, many variables which are of central importance in economics, e.g. the elasticities of commodity prices with respect to supply or demand, are not known with a precision better than 30% or 50%.[1] Other figures regarding the accuracy of economic data can be found in Morgenstern (1950). Raising the signal to noise ratio is a crucial challenge for the social sciences. In this chapter we describe three methods for improving signal identification and we illustrate them through specific social phenomena.

Before we begin, an additional remark is in order. Signal detection is an important topic in mathematical statistics. Here however, we propose upstream solutions to be used in the design phase of an experiment. Once the data have been recorded the fate of the battle against noise is largely settled. Using one statistical technique rather than another will improve matters only marginally.

3.1 The extreme value technique

Suppose we wish to study how the period of a pendulum depends upon the initial angular deviation θ_0. If we try initial amplitudes of 5, 10, 15, 20, 25 and 30 degrees the corresponding periods will differ by less than 5%. Thus, unless our measurements are very precise it will not be clear if the period depends upon θ_0

---

[1] The fact that the elasticities are fluctuating in the course of time shows that many *different* phenomena are at work simultaneously. One would face the same situation when observing the movements of a pendulum in a train; instead of revealing the characteristics of the pendulum, the observations would largely reflect the curves and bends of the railroad line. The prime importance of the stochastic character of economic phenomena has been emphasized and studied by Masanao Aoki in several books the most recent of which is Aoki and Yoshikawa (2006).

or not. On the contrary, if we try initial angles of 10, 90 and 179 degrees the corresponding periods will be sufficiently different to show unambiguously that the period increases with θ_0. This, in a nutshell, is the rationale of the extreme value technique. We will now show how it can be used in social phenomena.

It has been known since the late nineteenth century that the suicide rate is higher for unmarried than for married people. The ratio is about three for men and two for women (more details will be given in a subsequent chapter). Obviously, in a population in which there are more men than women, not all men will be able to get married and one would therefore expect a higher suicide rate than in a population whose sex ratio is closer to one. Can this prediction be tested?

For most populations the sex ratio is confined in a narrow interval around one, usually between 0.95 and 1.05. In such a population the expected effect will be very small. If the suicide rate of married men is s, the suicide rate of males in a population with a sex ratio male/female of 1.05 will be $(95s + 5 \times 3s)/100 = 1.15\,s$.[2] For this effect to be detectable the background noise should be substantially smaller than 15%, which is not the case. Thus, the effect will not be observable in this way.

The challenge is to find populations whose sex ratio is strongly different from one. It is well known that immigrants usually have a gender proportion involving more males than females. For instance:

- In 1890 the sex ratio of Chinese immigrants in the United States was 27.
- In 1901 the sex ratio of Chinese immigrants in Australia was 75.

Of course, there are two additional requirements: (i) these populations must be large enough to produce a sizable number of suicides; and (ii) the suicide rates for these specific populations must have been statistically recorded. In answer to the first question, there were 107,000 Chinese people in the United States in 1890 and 30,000 in Australia in 1901. Although not very large, these populations are sufficient to produce fairly stable suicide rates, especially if one performs averages over several successive years. This leads us to the second question: are there available data for these populations? The yearly *Mortality Statistics* volumes give suicide numbers for Chinese immigrants in the United States but these data were published only after 1923. By this time the sex ratio had dropped to 6.5, a level which, fortunately, is still high enough to produce an observable effect. Figure 3.1 shows the suicide rate of people of Chinese descent in the continental part of the United States over the period 1923–60, a time interval during which the sex ratio fell steadily from 6.5 to about 1. As expected, the suicide rate decreases along with the sex ratio. Over this time interval of 37 years the suicide rate was divided by 3. Compared to such a big change, the fluctuations of 10% to 20% due to the background noise are of little importance. The

---

[2] For the sake of simplicity we assume that the male to female ratio is the only limiting factor on the number of marriages.

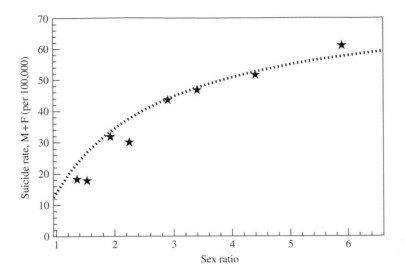

Fig. 3.1 Suicide rate in the Chinese community of the United States, 1923–1960. Horizontal scale: male to female sex ratio; the high ratios on the right-hand side correspond to the 1920s, while the low ratios on the left-hand side correspond to the 1950s. The data refer to the continental United States, which means that Hawaii (where there is also a substantial Chinese community) is excluded. As there were only about 30 suicides annually, we performed an average over two successive years. The suicide rate decreases with the sex ratio in the expected way (more details can be found in a subsequent chapter). It can be noted that: (i) The rest of the United States did *not* experience a steady decline in suicide rate; for instance there was a strong increase between 1923 and 1932 and a smaller increase between 1945 and 1950. (ii) Most of the Chinese people who were in the United States in the early 1920s had been there for several decades, as can be seen from the fact that the Chinese population in the United States reached a maximum in 1890 (107,000) and decreased steadily in subsequent decades until after 1940. *Sources: Mortality Statistics, annual reports 1923–37; various years; Vital Statistics of the United States, 1938–60, various years. These volumes are available online on the website of the National Center for Health Statistics.*

Chinese immigrants are the parallel of the deviation of 179 degrees in the pendulum experiment. Of course, for social phenomena the success of the method depends on data availability, but thanks to the Internet revolution the availability of statistical data has improved tremendously.[3]

The next section explains how signal identification can be improved if one has some knowledge about the date of occurrence, magnitude or shape of the expected signal.

---

[3] For instance, all the volumes in the series of *Mortality Statistics* and the subsequent (after 1938) series of *Vital Statistics of the United States* are available on the website of the National Center for Health Statistics. More details on this question will be given in a subsequent chapter. Incidentally, it can be noted that we have not yet been able to find suicide data for Chinese immigrants in Australia.

3.2 Pattern matching: knowing when and what to observe

It turns out that young widowers under the age of 30 experience suicide rates which are almost 10 times higher than in the rest of the population. However, we do not know what is the average time interval η between the deaths of the wives and the suicide occurrences. The only information that we have is that η is less than three years. It would be very useful to know if η is of the order of one week, one month or one year. There is a similar uncertainty about the time constant of many economic mechanisms. For instance, the theory of international trade tells us that, when a country has a permanent trade and current account deficit, the exchange rate of its currency should fall until the deficits are brought down. However, we do not know whether this is supposed to happen after one, five or ten years.[4] Not knowing the time lags puts us in a fairly awkward position. To point out how weird such a situation would appear in physics, let us return to the parallel with the pendulum. Suppose that the pendulum is at rest in the vertical position and that an impulsive force is applied to its mass. The existence of a time lag would mean that the mass begins to move only after, say, a few seconds or minutes; this would seem very surprising and clearly shows that lagged responses are fairly uncommon in physics. In this section we present an example in which the reaction time of the system is well known; as a result one knows precisely at what point in time the signal will appear. It will be seen that this knowledge greatly improves signal identification.

The population pyramid of Japan based on the census of 2000 presents a mysterious discontinuity for people aged 34, that is to say who were born in 1966 (Fig. 3.2c) The number of people (both males and females) born in this year is much smaller than in 1965 or in 1967. The difference is of the order of 30%. Further investigation reveals that 1966 was a Hinoeuma year, which means a Fire Horse year in the Chinese calendar sometimes used in Japan.[5] Girls born in that year grow up to be known as "Fire Horse women" and are reputed to be headstrong and to bring bad luck to their families and to their husbands. In 1966, as a baby's sex could not be reliably identified before birth, there was a big increase in the number of abortions, which brought about the sharp fall in birth rate observed on the population pyramid. According to the Chinese calendar, Fire

---

[4] For instance, Australia has had a trade and current account deficit for several years in the early 2000s, but at the time of writing (June 2006) these deficits had not brought about a fall in the exchange rate of the Australian dollar. On the contrary, between January 2002 and January 2006, the Australian dollar progressed against the US dollar (from 0.52 to 0.75) as well as against the euro (from 0.58 to 0.61). One should add that interest rates in Australia were not substantially higher than in the United States. Even if the exchange rate of the Australian dollar eventually falls it will be difficult to say if it was indeed the deficit which was the crucial factor. In a general way, the longer the time lag beween cause and effect, the more difficult it is to demonstrate the existence of a causal link (because of the larger number of exogenous shocks).

[5] Because the Chinese New Year occurs in late January or early February, the Fire Horse year does not exactly coincide with 1966; in fact, it started on January 21, 1966 and ended on February 8, 1967.

Horse years occur every 60 years; thus, the three previous one were in 1906, 1846 and 1786. It is a natural question to see if the effects in those years were similar to the one in 1966.

Figure 3.2a shows the curve of the male/female sex ratio in 1888, which is the first year for which such data are available. We see a distinct spike which corresponds to people aged 42, that is to say born in $1888 - 42 = 1846$. It corresponds to a sex ratio about 15% greater than its normal level of about 1.05.[6] The lower curve shows the total, male plus female, population. One would expect a dip of (at least) $15\%/2 = 7.5\%$; it happens to be somewhat larger at 11%.

Is it possible to detect an after-effect due to the Fire Horse year of 1786? People born in this year would be $1888 - 1786 = 102$ years old in 1888; unfortunately there are no data available for age groups over 84 years, which means that no

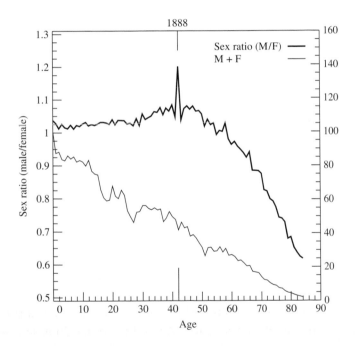

Fig. 3.2a Sex ratio and population by age in Japan (1888). In this graph (as well as in Figs. 3.2b, c) the thick line corresponds to the male/female sex ratio (left-hand side scale), the thin line to the total population normalized to 100 at age 0 (right-hand scale) and the thin vertical lines indicate the cohorts born in a Fire Horse year. Age 0 means aged less than one year. For the age group born during the Fire Horse year of 1846 there is a 15% increase in the sex ratio and an 11% fall in the total population. *Sources: Same as Table 3.1.*

---

[6] This corresponds approximately to a deficit of 38,000 girls.

direct observation is possible.[7] However, if there was a deficit in girls of the same magnitude as in 1846, this should have led to a reduction in the number of marriages about 20 years later; thus, the generation born around $1786 + 20 = 1806$ should be somewhat smaller. In 1888, the people born in 1806 are 82 years old; can we identify an indentation in the total population curve in the vicinity of 82 years? The answer is no. Even if the last part of the curve is greatly magnified no trough can be detected. It is probable that the dip which may have existed at birth was fairly broad and was further leveled off and smoothed during the 82 years of life of these people.[8]

After 1786 and 1846 we come to 1906. This effect is described in Fig. 3.2b. This figure gives the population pyramid in 1913. There is a spike for people

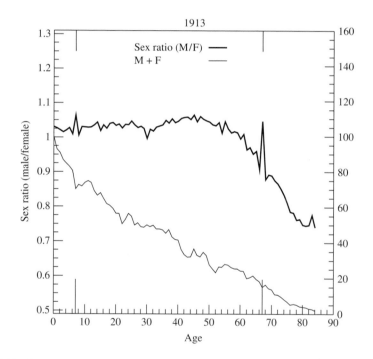

Fig. 3.2b Sex ratio and population by age in Japan (1913). For the cohort born in 1906 the sex ratio spike has an amplitude of 4% and the fall in population is 11%. *Sources: Same as Table 3.1.*

---

[7] The fact that the sex ratio spikes have a good persistence in time is shown by the population pyramid of 1925 in which the cohort born in 1846 is 79 years old; the spike in the sex ratio is still clearly visible even though it has been somewhat eroded by the higher male than female mortality that prevails in old age.

[8] Naturally one could try to repeat this reasoning a second time. As the cohort born in 1806 was reduced, the cohort born 20 years later should also have been smaller. These people would be 62 years old in 1888. There is indeed a small indentation for this age group. However, it would be hazardous to draw any definite conclusion. For one thing our argument rests on the assumption that all people get married and have their first child at the age of 20 which is of course a rough approximation.

who are 7 years old, that is to say who are born in $1913 - 7 = 1906$. The sex ratio spike of people born in 1906 has an amplitude of 4% and the trough in the male plus female curve has an amplitude of 11%. As statistical birth data are available for 1906, it is possible to check our previous conclusions. It turns out that the total number of births in 1906 is 11% lower than the average for the other years of the decade 1901–10. The sex ratio at birth is 4.3% higher than the average of the decade. Thus, these figures indeed confirm those that we read on the population pyramid. For people who are 67 years old (i.e. born in 1846) the sex ratio spike is still clearly visible but the small trough in the male plus female curve would be indiscernible if one did not know where to look for it.

After 1906 we come to 1966. The population pyramid of 2000 (Fig. 3.2c) shows the sharp trough already evoked. One may wonder if, as in 1846 and 1906, there was an anomalous sex ratio at birth. Of course, one expects it to be much smaller than in 1906 and this is already visible on the sex ratio curve. More detailed birth data show that in 1966 the sex ratio was 1.3% larger than over the other years of the decade 1961–70. This small excess sex ratio can no longer be detected in the population pyramid of 2000 when the Fire Horse generation is 34 years old.

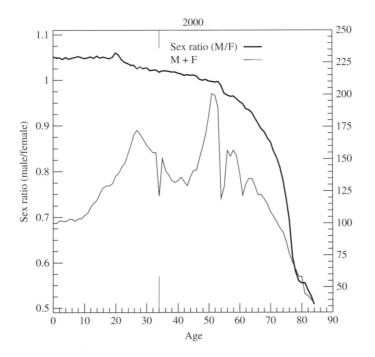

Fig. 3.2c Sex ratio and population by age in Japan (2000). The Fire Horse year of 1966 led to a sharp trough in births but, in contrast to 1846 and 1906, the sex ratio remained almost normal; more precise figures are given in Table 3.1. *Sources: Same as Table 3.1.*

Table 3.1 *Male and female births in three Fire Horse years*

	1846	1906	1966
Total births (%)	−11	−11	−24
Female births (%)	−19	−13	−24
Sex ratio (%)	+20	+4.3	+1.3

Notes: The percentages refer to the differences between the year under consideration and the mean of the nine other years in the same decade. For 1846, as no birth statistics are available, the percentages were derived from the population pyramid of 1888. Surprisingly, the 1966 decrease in total births was substantially higher than in previous Fire Horse years. One may wonder if this should not be attributed to the influence of an additional factor. A possible candidate is the after-effect, one generation later, of the fall in birth rate which occurred in 1946.
Sources: Historical Statistics of Japan (http:// www.stat.go.jp): Population by single years of age and sex, live births by sex and sex ratio of live births; Matsumoto (1975).

The sex ratios at birth for the three Fire Horse years for which data are available are summarized in Table 3.1. There is a last question that Fig. 3.2c can help to answer: what effect does a sudden drop (or increase) in birth rate have one generation later? In accordance with the extreme value technique exposed in the previous section, we first examine the effect of the huge peak (*P*) that occurred between 1947 and 1952 (visible on Fig. 3.2c, ages between 48 and 53). There is indeed a subsequent peak (*P'*) in the male plus female curve around the age of 27. Three additional observations can be made.

- The magnitude of *P'* is only half the magnitude of *P*.
- *P'* is about twice as wide as *P*.
- The time lag between *P* and *P'* is about 25 years.

Can we make a similar observation for sharp troughs? The answer seems to be yes. There is a sharp trough at age 54 that corresponds to people born in 1946; these people were 20 years old in 1966, that is to say in coincidence with the Fire Horse year. Therefore, it is quite plausible that a fraction of the birth rate trough of 1966 should be attributed to the after-effect of the birth rate trough of 1946.

In conclusion, we have seen that the 60-year periodicity in the occurrence of the Fire Horse years was of considerable value in helping us to decipher the fluctuations in the population pyramids. In this specific case, we relied on a recurring time pattern, but it is clear that any pattern, whether in time, space or any other variable will be of great usefulness.

We now turn to a third method of signal identification which can be seen as an extension of the law of large numbers.

3.3 Reducing noise by adding up several realizations

Suppose that for the purpose of a classroom experiment one wishes to measure the period T of a pendulum. A standard procedure is to measure the time t_{10} of 10 oscillations and get the period by dividing by 10: $T = t_{10}/10$. Such a procedure makes sense because the main uncertainty comes from the operation of starting and stopping the chronometer. If γ designates each of these uncertainties, the relative error on the measurement of T will be $2\gamma/t_{10} = (1/10)(2\gamma/T)$, whereas it would be $2\gamma/T$ if only one period had been measured. The only reason why we made this rather trivial point is that it shows that in itself the rationale of the procedure has little to do with the law of large numbers. It is only if we assume that all periods are in fact slightly different (due to vibrations, friction, draft) that the addition of random variables will play a role. As one knows, the argument relies on the two following rules.

- The variance of a sum of two independent random variables X, Y is the sum of the variances: $\sigma^2(X + Y) = \sigma^2(X) + \sigma^2(Y)$.
- The variance of the variable λX is given by $\sigma^2(\lambda X) = \lambda^2 \sigma^2$.

Combining these results and assuming in addition that X and Y have the same standard deviation σ one gets

$$\sigma^2\left(\frac{X+Y}{2}\right) = 2\frac{\sigma^2}{2^2} = \frac{\sigma^2}{2}$$

Similarly for the average of 10 random variables one gets

$$\sigma\left(\frac{1}{10}\sum_{i=1}^{10} X_i\right) = \frac{1}{\sqrt{10}}\sigma \simeq \frac{1}{3.2}\sigma$$

Thus by measuring 10 vibrations the measurement will be roughly three times more accurate.[9] Unfortunately, this argument is of little usefulness in the social sciences for two main reasons: (i) usually, the different realizations are not independent; (ii) very often one cannot repeat the experiment as often as one would like. Let us illustrate these difficulties by a few examples.

In the previous chapter we mentioned that the Werther experiment can be easily repeated. Obviously, however, there is a limiting factor which is the number of suicides announced on the front page of the *New York Times*. In the 1950s and 1960s there were on average between one and two suicides every year which were announced on the front page. Thus, over a period of 20 years the number of realizations will be of the order of 40; the square root of 40 is 6.3. It remains to be seen if this factor is large enough. We come back to this point later on.

---

[9] Naturally, this argument does not apply if the system is not stationary, for instance if a window has been left open which provokes an increasing amount of draft.

Sometimes data are available for a large number of realizations but they are not independent. For instance, in 2004 there were 2,768 stocks and 1,059 bonds listed on the New York Stock Exchange. For each of these stocks, data are available in the form of time series. Unfortunately, these time series are not independent. This is fairly obvious for companies which belong to the same economic sector. Thus, in 2004–5 the stock prices of Exxon Mobil and of Chevron (two companies involved in the production of oil) were highly correlated with a correlation of about 0.95. Stock prices of companies in different economic sectors may be correlated as well because some variables (e.g. the number of mergers and acquisitions, the number of buybacks, the inflation rate) affect almost all stocks globally. As is fairly obvious intuitively (and will be shown later on), when two time series are highly correlated taking their average does not reduce the standard deviation.

This discussion suggests that in order to be helpful in the social sciences the standard argument must be extended in two directions.

- Because the number n of realizations is often limited, one would like the standard deviation of the average to decrease faster than $1/\sqrt{n}$.
- Because the realizations are often interdependent, the probabilistic argument should be extended to include correlated random variables.

As will be seen, the second requirement will help us to fulfill the first condition as well.

For the sake of simplicity we consider the average of a sum of three correlated random variables X_1, X_2, X_3 of mean zero and identical standard deviation σ. The conditions on the mean and standard deviation are not a limitation because if initially the variables X_i' do not satisfy them it is always possible to carry out the transformation $X_i = (X_i' - m_i')/(\sigma_i'/\sigma)$ where m_i', σ_i' denote the mean and standard deviation of X_i'. Our objective is to compute the standard deviation of

$$Y_3 = S_3/3 \qquad S_3 = X_1 + X_2 + X_3$$

In accordance with the rules stated previously,

$$\sigma^2(Y_3) = (1/3^2)\sigma^2(S_3)$$

By definition of the variance and due to the fact that the expectation of S_3 is equal to 0,[10] one gets

$$\sigma^2(S_3) = E\left[S_3^2 - E^2(S_3)\right] = E\left[\sum_{i=1}^{3} X_i^2 + 2(X_2X_3 + X_3X_1 + X_1X_2)\right]$$

$$= \sum_{i=1}^{3} E(X_i^2) + 2\left[E(X_2X_3) + E(X_3X_1) + E(X_1X_2)\right]$$

---

[10] It should be recalled that the expectation of a sum of random variables is always equal to the sum of the expectations whether the variables are independent or not.

We express the expectations of the products by introducing the coefficient of correlation of the X_i:

$$r_{12} = \frac{E\left[(X_1 - E(X_1))(X_2 - E(X_2))\right]}{\sigma(X_1)\sigma(X_2)} = \frac{E(X_1 X_2)}{\sigma^2}$$

Thus

$$\sigma^2(S_3) = 3\sigma^2 + 2\sigma^2(r_{23} + r_{31} + r_{12})$$

Introducing the mean of the r_{ij}, $\bar{r} = (r_{23} + r_{31} + r_{12})/3$, we obtain

$$\sigma^2(S_3) = 3\sigma^2[1 + 2\bar{r}]$$

and finally

$$\sigma(Y_3) = \frac{\sigma}{\sqrt{3}}\sqrt{1 + 2\bar{r}} \tag{3.1a}$$

This formula has an obvious generalization to an arbitrary number n of random variables:

$$\sigma(Y_n) = \frac{\sigma}{\sqrt{n}}g \qquad g = \sqrt{1 + (n-1)\bar{r}} \tag{3.1b}$$

where

$$\bar{r} = \frac{1}{[n(n-1)/2]}\sum_{i \neq j}^{n} r_{ij}$$

When the r_{ij} are all equal to zero $g = 1$ and we get the standard result for independent variables. On the other hand, $\bar{r} = 1$ implies that all the r_{ij} are equal to 1; in this case the three variables are identical (with probability 1) and one gets $\sigma(Y_3) = \sigma(X_1) = \sigma$ in agreement with Formula (3.1).

Formula (3.1) has interesting implications when $\bar{r}$ becomes negative. First we observe that for $n = 3$ the smallest value that $\bar{r}$ can take is $\bar{r} = -1/2$. In other words, for three random variables, it is impossible that $r_{23} = r_{31} = r_{12} = -1$. This makes sense intuitively because $r_{12} = -1$ and $r_{13} = -1$ imply that $X_2 = -X_1$ and $X_3 = -X_1$, which implies of course that $X_2 = X_3$ and $r_{23} = 1$, hence $\bar{r} = (-1 - 1 + 1)/3 = -1/3$.

When $\bar{r}$ is equal to -0.5 the standard deviation of Y_3 is equal to zero. In other words, the background noise represented by σ is completely eliminated. Just to show how dramatic this effect can be, we set up a simulation in which a small deterministic signal, a lightly damped vibration, has been added to white noise. As the amplitude of the deterministic signal is only a fraction of the amplitude of the noise, it is completely hidden, as can be seen in the first line of Fig. 3.3. The two series have zero mean, but they do not have the same standard deviation: $\sigma_1 = 0.97$ and $\sigma_2 = 0.49$. The correlation of the two series is -0.96. The average of the two series (bottom left) is almost as noisy as the initial series and the deterministic signal

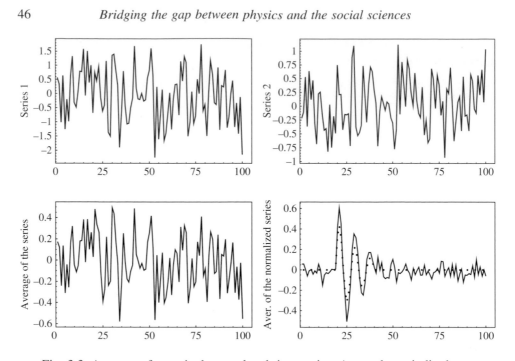

Fig. 3.3 Average of negatively correlated time series. A pseudo-periodic deterministic signal is hidden in series 1 and 2; the challenge is to extract it. The two series have zero mean, correlation of -0.96 and standard deviations of 0.97 and 0.49 respectively. Taking the average of the two series (bottom left) reduces the level of noise but not enough to make the deterministic signal clearly visible. However, if we take the average after dividing the series by their standard deviations, one expects the level of noise to be cut drastically. This is indeed what happens: the averaging process reveals the deterministic signal almost in its initial shape (which is represented by the dotted curve). It can be noted that the technique of computing the autocorrelation function which is often used to reveal hidden periodicities does not work when applied to the bottom left graph: the pseudo-periodic component is just too small. More generally the same procedure works for n series containing a common deterministic component provided that their average cross-correlation is close to $-1/(n-1)$.

is still invisible. However, if we normalize the series by dividing them by their standard deviations before taking their average, the level of noise is drastically reduced and the deterministic signal becomes clearly visible. Of course, this is a simulation and such a dramatic effect is not likely to be observed with real series. Nonetheless, we will show by two illustrations that the method can indeed be helpful.

Application 1 We build a data set of daily stock prices for 24 stocks: 20 stocks from the Dow Jones Industrial plus the first 4 stocks of the Standard and Poor's 500 sample. All series cover the period from January 1999 to the end of 2002. As these series are mostly for large corporations, they follow more or less the price evolution of the market as described for instance by the evolution of the Standard and Poor's 500 index. In order to get rid of this common trend, we divided the series by the S&P500 index. Then, we normalized the series to reduce their mean to zero and their standard deviation to 1. With these 24 series, $(24 \times 23)/2 = 276$ pairs can be formed; for each of these pairs we computed the correlation coefficient.

- It turns out that the pair which gives the most negative correlation is (Hewlett-Packard, Altria).[11] The correlation is -0.84, which gives a coefficient g equal to $\sqrt{1-0.84} = 0.40$; this means that, due to the negative correlation, the standard deviation is divided by 2.5 with respect to what would be obtained with two independent time series.
- With the 24 series it is possible to form $(24 \times 23 \times 22)/(1 \times 2 \times 3) = 2,024$ triplets. For each of these triplets (i, j, k) one can compute the average $\bar{r}$ of the three cross-correlations r_{jk}, r_{ki}, r_{ij}. The triplet which turns out to have the most negative $\bar{r}$ ($\bar{r} = -0.40$) is (American International Group, Du Pont de Nemours, Hewlett-Packard).[12] In this case $g = 0.45$, which means that the standard deviation is 2.2 times smaller than the standard deviation of three non-correlated series.
- The same operation can be performed for the 10,626 quadruplets, the 42,504 quintuplets, the 134,596 sextuplets, etc. The only practical limitation is the computing time which rises very rapidly. Thus, it took about 40 hours of computing time to identify the quintuplet which gives the most negative correlation. It corresponds to the following companies: Du Pont, Hewlett-Packard, Intel, Altria, Merck.[13] The average correlation is $\bar{r} = -0.21$, which gives $g = 0.40$. It would require $5/0.40^2 = 31$ non-correlated companies to achieve the same reduction in standard deviation. The main advantage of achieving the same reduction with only 5 companies lies in the fact that a common deterministic signal is more likely to be found in 5 companies than in 31.

Of course, once noise reduction has been carried out, the main question is how to interpret the resulting average. For instance, in the 200 days after the shock of September 11, 2001, the average of the previous five stocks displays a succession of sinusoidal oscillations of increasing period and increasing amplitude:

$$(25 \text{ days}, 0.06) \quad (45 \text{ days}, 0.10) \quad (77 \text{ days}, 0.14) \quad (92 \text{ days}, 0.20)$$

This kind of pattern may give us an insight into the main vibrating modes of the New York stock market, but it must of course be confirmed by the observation of a similar pattern in the wake of other major shocks.

[11] Hewlett-Packard is a computer company, while Altria is (since 2003) the new name of the Philip Morris company.
[12] American International Group is a consortium of insurance companies, Du Pont is a chemical company.
[13] Intel is a manufacturer of semiconductor chips; Merck is a pharmaceutical company.

Application 2 This second application is not strictly speaking about signal iden-tification, but it has much to do with noise, addition of random variables and intercorrelations. It will help us gain an understanding of what determines the standard deviation of suicide rates at county, state and nation level. This knowl-edge will be of great importance for the discussions of the Werther effect at the end of the chapter.

First, we describe the procedure which leads to the curves in Fig. 3.4. We selected (fairly randomly) six counties whose populations comprise between 80,000 and 100,000 inhabitants. The average population of these counties, namely 0.09 million, defines the x-value in the graph of Fig. 3.4. From the database of the Centers for Disease Control, we get the number of suicides in each year of the 20-year long time interval 1979–98. This allows us to compute the standard deviations of the six time series; their average, namely $\sigma = 3.76$, defines the y-coordinate of the square in the graph of Fig. 3.4. In order to estimate how

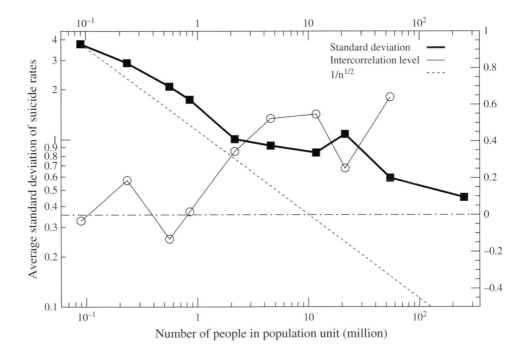

Fig. 3.4 Standard deviation of suicide rates in the United States. The graph represents the standard deviation of suicide rates as a function of the population n of population units of increasing size. The intercorrelation curve (right-hand scale) represents the average intercorrelation of different units in the same size group. The broken curve (scale not shown) represents the function $y = 1/\sqrt{n}$. A possible reason explaining why the standard deviation breaks away from the curve $1/\sqrt{n}$ is explained in the text. *Source: WONDER database on the website of the Centers for Disease Control.*

much these series are interdependent, we compute the pair correlations of the 15 pairs that can be formed with the six series. The average of these correlations, namely -0.03, defines the y-value of the circle (the corresponding scale is on the right-hand side). Then, we repeat this procedure for population units of increasing sizes. The last square on the far right corresponds to the United States. No correlation can be computed in this case because there are no other series in this size group. While the two solid curves correspond to observations, the dashed line shows the theoretical function $y = 1/\sqrt{x}$ that would be expected if all series were uncorrelated.

How should Fig. 3.4 be interpreted? It shows that the empirical curve remains close to the $1/\sqrt{x}$ curve until the population reaches a threshold size of about one million. For population units over one million, the decrease in the standard deviation is slower than $1/\sqrt{x}$ and at the same time the intercorrelations increase. At this point we must say a little bit more about the curve $y = 1/\sqrt{x}$. Unlike stock prices for which there are no standard models, suicide rates can be described in a natural way as the increment of a Poisson process. A Poisson process $X(t)$ is defined by the assumption that during each time interval Δt there is a probability $\lambda \Delta t$ that a new suicide occurs. This process models the cumulative number of suicides, which is an increasing function of time. The number of suicides in a given time interval is, one might say, the derivative of the Poisson process; for instance the annual number of suicides $Y(t)$ is defined as the increment of $X(t)$:

$$Y(t) = [X(t + \theta) - X(t)]/\theta$$

where θ represents one year. It can be shown that, according to this simple model, the standard deviation of the suicide rates is given by $\sigma = \sqrt{t_m/N\theta}$, where t_m denotes the average suicide rate and N the size of the population unit under consideration (Papoulis 1965, p. 287). The fact that σ is proportional to $1/\sqrt{N}$, as in the addition of independent random variables, does not come as a surprise because it is known that a Poisson process can alternatively be defined as a sum of an increasing number of random variables (Papoulis 1965, p. 558). Why then does the curve of the standard deviation break away from $1/\sqrt{x}$ in the population range over one million?

A possible interpretation consists in assuming that there are two different sources of noise which affect suicide rates. (i) A purely random component which leads to a standard deviation σ_r and is well described by the Poisson increment process; this random component corresponds to a large number of factors at individual or local level. (ii) A deterministic source of noise which leads to a standard deviation σ_d and which corresponds to the response of the system to a few macro-factors at regional or national level. As examples of such macro-factors one can mention the marriage and divorce rates or the unemployment rate.[14] In short, $\sigma = \sigma_r + \sigma_d$, $\sigma_r \sim 1/\sqrt{N}$, $\sigma_d \simeq 0.5$.

Of course, the σ_d component exists at all population levels, but for small units σ_r is large enough to make σ_d almost invisible in relative terms.

The nature of the deterministic factors can be confirmed by the correlation curve. Suppose for a moment that these factors are local factors at county level. In this case there would be no reason for the intercorrelation of suicides to increase with unit size. In contrast, if the deterministic factors are indeed macro-factors at regional or national level, it is not surprising that they may bring about increased intercorrelation.[15]

We now present a brief discussion of the questions of statistical significance and confidence intervals.

3.4 Confidence intervals and statistical significance

The concepts of signal to noise ratio and of error bars that we used in this chapter are commonly used in physics and in electrical engineering. Econometricians as well as the social scientists who have adopted the language of econometrics rely rather on the notion of test of significance. In this section we give particulars about the notions of confidence intervals and test of significance and how they are related.[16] First we will recall the definition of statistical significance; then we give some typical orders of magnitude; finally, we discuss relevant applications to the social sciences. However, before we begin, it should be recalled that when the signal to noise ratio is high, say higher than 10, the results are clear-cut no matter what statistical tests are used to analyze them; on the contrary, if the signal to noise ratio is low, the conclusions will be uncertain unless one has additional information about the expected signal, which allows pattern matching (or similar procedures) to be used.

If a random variable X has a Gaussian frequency distribution of mean m (which we take equal to zero for the sake of simplicity), and standard deviation σ, X will

---

[14] Empirical studies show that the correlation between unemployment and suicide is fairly low but there can be an indirect relationship. A possible connection may be through the disruption of the family unit occasioned by unemployment. If this interpretation is correct, the connection between unemployment and suicide should be very dependent upon the level of social protection; thus, one would expect the connection to be lower in Scandinavian countries than elsewhere.

[15] The fact that marriage or unemployment rates are not necessarily the same in nearby counties is irrelevant for our argument. What matters is the rate of change of these factors over a period of 20 years. It turns out that these changes are slow and smooth enough to be called deterministic.

[16] To statisticians and econometricians the approach that we use may appear rather unsophisticated. However, one should keep in mind in a general way that the more sophisticated a statistical test, the larger the set of assumptions on which it relies and the greater the difficulty of checking that the data set under consideration indeed satisfies these requirements. Very often this proof is simply omitted either because the data set is too limited to permit the appropriate verifications or because it would be too time consuming to do so.

fall into the interval $(-1.96\sigma, 1.96\sigma)$ in 95% of the drawings (Ventsel 1973, p. 307). Equivalently, we can say that there is a likelihood of 0.05 for random drawings of $|X|$ to be larger than 1.96σ. Thus, if a spike that one believes to be a signal has an amplitude around 1.96σ, where σ is the standard deviation of the background noise, there is one chance in 20 that the "signal" is in fact a random noise fluctuation.[17]

Naturally, if the amplitude of the signal is larger than 1.96σ one would expect the likelihood that it is noise to be smaller. Table 3.2 summarizes some typical values. It should be emphasized that from a scientific perspective there is no threshold of significance that is better than another. In other words, a probability threshold of 0.1 is just as "good" as one of 0.001. Whether one adopts one particular threshold or another is an issue which depends upon the context of the experiment. If the number of events produced by the experiment is large one can be more selective, if the number of events is small, being very selective would make most results non-significant and would therefore be a fairly unproductive procedure. To make this discussion more concrete we consider an application to the specific case of the Werther effect.

Statistical significance of the Werther effect Is there really an increase in the suicide rate after the suicide of a celebrity or is the effect too small to be clearly detected? We examine how the previous notion can help us to answer this question. In his paper of 1974, Phillips analyzes $n = 33$ events. Each event is a suicide story published on the front page of the *New York Times*. In each case, Phillips compares two variables: (i) an observed number of suicides s and (ii) what he calls an expected number of suicides e. For definiteness, let us consider the case of Marilyn Monroe. She died on August 6, 1962, so s will be the number of suicides which occurred in the United States during the month of August 1962.[18] From the *Vital Statistics of the United States* (Vol. II, Part A, Section 1, p. 77) we learn that $s = 1838$. In July, there were only 1659 suicides, but this increase is of little significance because of the seasonal pattern described in the previous chapter. To make the test independent of the seasonal pattern, Phillips uses the following procedure. He computes the average number of suicides in August 1961 and August 1963, which he calls the expected number of suicides:

---

[17] Strictly speaking, 0.05 is the probability $P\{|X|/\sigma > 1.96\}$, but due to the very rapid decrease of the Gaussian function the probability that $|X|/\sigma$ is substantially larger than 1.96 is so small that in fact

$$P\{|X|/\sigma > 1.96\} \simeq P\{|X|/\sigma \simeq 1.96\}$$

[18] Clearly if the suicide of a celebrity occurs late in the month, it makes more sense to consider the suicides in the subsequent month. Phillips chose the 23rd day as the cut-off point; thus if Marilyn Monroe's death had occurred on August 24, one would rather consider the suicides that occurred in September 1962.

Table 3.2 *Probability that a signal is in fact a random fluctutation*

	Signal (s) to noise ratio: $x = \frac{s}{\sigma}$								
	1	2	3.3	4	5	6	7	8	10
$P\{X > x\sigma\}$, one chance in	6	40	2000	30 000	3.4×10^6	10^9	0.8×10^{12}	1.6×10^{15}	1.2×10^{23}
$P\{|X| > x\sigma\}$, one chance in	3	20	1000	15 000	1.7×10^6	0.5×10^9	0.4×10^{12}	0.8×10^{15}	0.6×10^{23}

Notes: The table gives the probability that a Gaussian random variable X of mean zero and standard deviation σ is greater than a given threshold $x\sigma$; the probability is expressed as one chance in n drawings, thus one chance in 40 (which corresponds to a probability of 1/40) means that a signal greater than 2σ will on average be observed once in 40 random drawings. The second line gives the same information for the variable $|X|$; it is identical to the first line except for the multiplication by a factor of 2; indeed

$$P\{|X| > x\sigma\} = P\{X < -x\sigma \cup X > x\sigma\} = P\{X < -x\sigma\} + P\{X > x\sigma\} = 2P\{X > x\sigma\}$$

In most applications the first line is of greater interest because one is interested in deviations of a well-defined sign. In particle physics, the conventional threshold of significance is 5σ; in the social sciences it is rather 2σ. These rules are nothing but conventions which are roughly in relation with the total number of events. In an experiment producing 10^{30} events it would make sense to take 10σ as the threshold of significance. The numbers in the table which are not given in standard textbook tables (e.g. Ventsel 1973) have been computed by using an asymptotic approximation of the complementary error function, namely $\sqrt{\pi/2}x\exp(x^2/2)$ where x denotes the signal to noise ratio.

$e = (1,579 + 1,801)/2 = 1,690$. The test variable is defined as the difference between s and e: $p = (s - e)/e = 8.8\%$.[19]

How significant is a value $p = 8.8\%$? The answer is given in Table 3.2 under the condition that the distribution of the numbers p is Gaussian, and provided we know the standard deviation of p. In order to determine the distribution of p we need far more years than just 1961, 1962 and 1963. How many years do we need? As each year gives 12 values of p, a sample of 5 years will give 60 values, which is sufficient to determine the standard deviation but is too small a sample to study the shape of the distribution. Such a determination requires at least 250 data points, that is to say 20 years. When the test is performed over the period 1935–60 (shown in Fig. 2.3) the distribution of p is indeed found to be reasonably

[19] Phillips's paper (1974, p. 344) incorrectly gives $e = 1640.5$, which leads to the higher value $p = 12.0$; probably the mistake comes from a confusion with the following line which contains exactly the same number 1640.5. However, this mistake does not substantially affect the overall conclusion for the whole sample of 33 suicides.

Table 3.3a *Is there a Diana–Werther effect? Observation of single events*

Name	Date of death	Observed suicides (s)	$s_{m,y-1}$	$s_{m,y+1}$	Expected suicides (e)	$p\left(\frac{s-e}{e}\right)$ (%)	$\sigma(p)$ (%)	Signal to noise ratio
United States								
Marilyn Monroe	August 6, 1962	1838	1579	1801	1690	8.8	4.92	1.79
James Forrestal	May 22, 1949							
$m =$ May		1549	1455	1600	1527.5	1.41	4.92	0.29
$m =$ June		1567	1410	1507	1458.5	7.44	4.92	1.51
UK								
Diana	August 31, 1997							
M + F		420	381	381	381	10	11	0.91
F		103	96	105	100.5	2.5	16	0.16
M		317	285	276	280.5	13	12	1.08
M,15–24		43	37	31	34	26	38	0.68

Notes: The third column of the table gives $s = s_{m,y}$, the monthly number of suicides (in month m of year y) in the month in which the death of the celebrity occurred. The sixth column gives the expected number of suicides $e = \left[s_{m,y-1} + s_{m,y+1}\right]/2$. The variable p represents the deseasonalized number of suicides. For a stationary time series the average of p over several years can be expected to be close to zero; this is indeed verified for the series under consideration; consequently, in a given month p quantifies the percentage of excess-suicides due to exceptional events. $\sigma(p)$ denotes the standard deviation of p in the time series under consideration. In the case of James Forrestal we have shown the data for two successive months because he committed suicide toward the end of May. The lines labelled F and M show the suicide figures for females and males respectively. In the United States the monthly suicide data do not make a distinction between males and females, but such data are available for the UK. It turns out that the Diana effect is much stronger for men than for women. The last line shows that p is even larger for young men, but in this case the smaller number of suicides results in a larger standard deviation with the consequence that the signal to noise is not much improved. The idea behind the Diana effect is that the death of a woman celebrity can act as a kind of "virtual widowhood" for some of her admirers; indeed it is known (see Bojanovsky 1980 and Chéruy 2002) that young widowers have a suicide rate ten times higher than married males of same age, especially in the first three months after the death of their wives.
Sources: US data: *Vital Statistics of the United States*, yearly volumes, Grove and Hetzel (1968); British data: personal communication from Ms. Anita Brook (UK Office for National Statistics) to whom I express my grateful thanks.

Gaussian with zero mean and with a standard deviation $\sigma(p)$ equal to 4.92%.[20] This leads to a ratio $p/\sigma(p) = 8.8/4.92 = 1.79$. By using a table of the Gaussian

---

[20] In fact $\sigma(p)$ was computed over the period 1945–60 during which the series is more stationary than during the whole period; over the whole period one gets $\sigma(p) = 5.74$, this higher estimate should certainly be attributed to the non-stationarity.

integral (e.g. Ventsel 1973, p. 543), we find that there is one chance in 28 that such a deviation is merely a random fluctuation.[21] In other words, fluctuations of a magnitude of 8.8% occur every 28 months, that is to say almost every two years.[22] As already mentioned, whether or not this signal should be considered as being "significant" is a subjective rather than a scientific question. From a scientific point of view the real question is whether the previous identification procedure can be improved. Three methods will be proposed, but before that we would like to discuss the second case mentioned in Table 3.3a.

The death of Diana, Princess of Wales, on August 31, 1997 was not a suicide but an accident. The results in Table 3.3a show that the p number, that is to say the deseasonalized number of suicides, in September 1997 reached 10%. As England has a smaller population than the United States, it is not surprising that the standard deviation $\sigma(p)$ is larger. As a result the signal to noise ratio $p/\sigma(p)$ is lower than in the Monroe case. Perhaps of greater interest is the fact that there seems to be a fundamental difference in the reactions of males and females. For men the signal to noise ratio is 1.1 whereas it is almost zero for women.

3.5 Upgrading statistical tests

We discuss three methods for improving the previous results.[23] In each case we will need additional data. This illustrates our previous statement that identification can be improved in a substantial way only by including additional information.

- The first method consists in observing the effects of several deaths instead of just one. It is the same method as when one measures ten swings of a pendulum instead of just one.
- The second method consists in observing the effect of a death in several places instead of just one. The detection of gravitational waves by several detectors located in different countries relies on the same approach.
- The third method consists in observing the effect of a death over several months instead of just one in order to apply a pattern matching procedure.

We now discuss each of these methods in more detail.

Several events This is the method that Phillips (1974) used in his paper. As we already mentioned, he selected a sample of 33 suicides publicized on the front

---

[21] This result is consistent with the first line of Table 3.2: $28 \in (6, 40)$.

[22] What makes the Monroe case important is the fact that it is the event in Phillips's list which leads to the highest number of excess-suicides. Thus, in line with the extreme value technique, it is reasonable to begin to investigate this event in some detail.

[23] In the literature one can find numerous procedures of peak identification. Many of them are context dependent, such as identification in radar detection or in astronomy. Many others use specific mathematical tools such as Fourier or wavelet analysis. In the present section, we focus on basic ideas rather than on specific techniques.

page of the *New York Times*. In each case he computed the variable $p = (s - e)/e$. The first question which arises is how many of the p values are positive. One finds that p is positive in 26 of the 33 cases, which represents a percentage of 79%. The second question is: "Are the p values close to zero or markedly different from zero?" One finds that their average is 2.51%.

Next comes the crucial question: "Does 2.51% represent a deviation which is significantly different from zero?" From the previous section we know that for each of the events the standard deviation of p is equal to 4.92%. If the 33 events are uncorrelated the standard deviation of their average will be

$$\sigma\left(\frac{1}{33}\sum_{i=1}^{33} p_i\right) = \sigma(p)/\sqrt{33} = 0.86$$

But are the events really uncorrelated? The answer is yes. The argument goes as follows. On average the time interval between two events is 20 years/33 = 7.3 months. The events will be uncorrelated if the autocorrelation function of p falls to zero in a time which is shorter than 7.3 months. It turns out that the autocorrelation function falls to zero within 2 or 3 months. Thus, the signal to noise ratio is 2.51/0.86 = 2.92. By using the formulas given in Table 3.2 we find that there is one chance in 526 for this fluctuation of the average to be merely a random fluctuation. If one recalls that in the case of Marilyn Monroe the result was one chance in 28, we see that we have been able to greatly improve the level of significance.

Several places In the second method one explores the effect of one event in different places. Until 1950 the *Vital Statistics of the United States* provided monthly suicide data not only for the whole country but also for each state. Can we use these data to improve the significance of the test? To take advantage of this additional information we must select a suicide that occurred before 1950. In Phillips's list there are only five events which occur before 1950. To be in the most favorable position we select the event which gives the largest p value (i.e. 1.44%), namely the suicide of James Forrestal, which occurred on May 22, 1949.[24] We consider the effect of this suicide in a sample of 17 states.[25] Table 3.3b shows that the average of the p values for the 17 states is 2.32%. This value must be compared with the standard deviation of the average of p over the 17 states. Each state is characterized by a specific $\sigma(p)$ and their mean is 27%.

---

[24] In Phillips (1974, p. 344) the expected number of suicides e is incorrectly given as 1493.5 instead of 1527.5; 1493.5 is identical with the figure in the line immediately below, which suggests that the mistake is probably of the same kind as in the Monroe case, i.e. a confusion between two successive lines.

[25] These states were selected through the following criterion: they are the most populous states which belong to the registration area in 1912.

Table 3.3b *Is there a Diana–Werther effect? Observation of several events or places*

		$p\left(\frac{s-e}{e}\right)$ (%)	$\sigma(p)$ (%)	Signal/noise ratio
Several events				
1	33 celebrities	2.44	0.86	2.84
Several places				
2	Forrestal, 17 states Diana, 3 regions	2.32	6.55	0.35
3	M	21	10	2.10
4	F	8.3	22	0.38

Notes: As in previous tables, p represents the deseasonalized monthly number of suicides, a variable whose average over several years is almost equal to zero. In the Diana case the three regions are England, Wales and Scotland. The comparison of cases 3 and 4 confirms that the effect is stronger for males than for females. The biggest improvements in the signal to noise ratio are cases 1 and 3.
Sources: Same as in Table 3.3a.

If the series are uncorrelated[26] the standard deviation of the average is obtained by computing the sum of their variances divided by the number of states, which gives approximately $27/\sqrt{17} = 6.5\%$. This result makes sense because it is of the same magnitude as the figure of 4.92% corresponding to the United States but slightly greater due to the fact that the 17 states are smaller than the United States. Thus, we get a signal to noise ratio equal to $2.32/6.5 = 0.35$. This figure should be compared with the result obtained for Forrestal at the level of the whole country, i.e. $1.44/4.92 = 0.29$. In other words, the signal to noise ratio was indeed improved, but only marginally. This is due to two circumstances. (i) In 9 of the 17 states p was negative, which shows that in a majority of the states the signal was smaller than the background noise. (ii) Furthermore, many states have relatively small populations and therefore their $\sigma(p)$ are fairly large. For instance, $\sigma(p)$ is equal to 11% in California but it becomes as high as 45% in Colorado whose population is seven times smaller.

In the previous attempt we used what can be called an indiscriminate, systematic statistical analysis. As we had no reason to expect the effect to be greater in one state than in another we treated them all in the same way. However, this is not the only possible strategy and in fact it may not be the most appropriate in the exploratory phase of the investigation. An alternative strategy is to adopt a case-study approach which focuses on specific states in order to discover underlying determinants. Let us illustrate this approach by the example of the suicide of

[26] When one computes the average intercorrelation over a sample of 10 pairs one gets indeed a correlation which is close to zero.

film star Carole Landis (second entry in Phillips's list, 1974, p. 344). Born in Wisconsin, she died in California on July 5, 1948, at age 29. At the global level of the United States, the *p* value of the suicides in July 1948 is positive but fairly small, $p = 1.72\%$. However, one may expect the impact of her death to be stronger in states in which she was well known, for instance Wisconsin or California. In Wisconsin *p* takes on a negative value, but in California *p* is not only positive but fairly large, $p = 18\%$. For a proper interpretation of this result one would need to know the amounts of media coverage of her career before her death and of her suicide in the month following her death.[27]

The same technique could be applied also to the death of Diana. In this case we can use separate data for England, Wales and Scotland. By the same reasoning we get a signal to noise ratio equal to 2.1 for males and 0.38 for females. For males the signal to noise ratio was doubled, which confirms that males reacted much more than females.

Pattern identification The rationale of the third method can be explained as follows. We do not yet have any knowledge of the time dependence of the Werther and Diana effects. However, if these effects really exist one would expect the number of excess-suicides to decrease progressively after the peak instead of falling abruptly to zero. In other words it should be possible, at least for the largest peaks, to detect an excess number of suicides not just in the month following the death of the celebrity but in several subsequent months. This is indeed what is observed in the Monroe and Diana cases (Table 3.4).

Intuitively, one would expect the occurrence of such a relaxation pattern to be relatively rare in a random series. The question is how rare is it exactly? The probability of such a pattern can be estimated through the following reasoning. We consider a time series Y_i of Gaussian white noise of mean zero and standard deviation σ. The expression "white noise" means that values at different times are uncorrelated, a property which can be checked by verifying that the autocorrelation function is almost equal to zero for all non-zero time lags. For definiteness we consider a pattern for which

$$Y_i \geq 2\sigma \quad \text{and} \quad Y_{i+1} \geq \sigma \qquad (3.2)$$

What is the probability of such a pattern? From Table 3.2 we know that

$$P\{Y \geq 2\sigma\} = 1/40 \quad \text{and} \quad P\{Y \geq \sigma\} = 1/6$$

If the time series has 10^6 points, the first assertion means that $10^6/40 = 25{,}000$ points will be above the 2σ level; we call these points 2σ-points. Now consider

---

[27] An approach to the Werther effect based on the amount of media coverage was tried by Stack (1987).

Table 3.4 *Excess-suicides in the months following the event*

	Month −1 (%)	Month 0 (%)	Month 1 (%)	Month 2 (%)	Month 3 (%)
p					
Monroe, M+F (%)	−2.5	8.8	7.0	3.8	2.5
Diana, M (%)	−11	13	5.9	16	21
$p/\sigma(p)$					
Monroe, M+F (%)	0.51	1.79	1.42	0.77	0.51
Diana, M (%)	−0.92	1.1	0.49	1.33	1.75

Notes: The table gives the deseasonalized number of suicides $p = (s-e)/e$ and the ratios $p/\sigma(p)$ in the months before and after the death of a celebrity. Month 0 is the month in which the death of the celebrity occurs. Month −1, the month preceding the death, is shown for the purpose of verifying that there is no overall trend. The fact that in the Diana case p increases in months 2 and 3 is probably due to an as yet unidentified exogenous shock.
Sources: Same as in Table 3.3a.

the points which immediately follow the 2σ-points. They constitute a subsample of points whose values are independent of the 2σ-points because of the white noise assumption. Thus, we can apply the same reasoning to this subsample, which leads to the result that there are

$$\left(10^6 \times \frac{1}{40} \times\right) \frac{1}{6} = 4{,}166$$

points which fulfill the requirement (3.2). This result (which can easily be checked by running a simulation) is summarized in the following rule:[28]

Identification of relaxation patterns Y_i is a Gaussian white noise time series of mean zero and standard deviation σ. We consider the probability $P\{R\}$ of observing a relaxation pattern described by the following event:

$$R = \left\{ \frac{Y_i}{\sigma} \geq a_0, \frac{Y_{i+1}}{\sigma} \geq a_1, \ldots, \frac{Y_{i+n}}{\sigma} \geq a_n \right\} \qquad a_0, a_1, \ldots, a_n > 0$$

$P\{R\}$ is given by the following formula:

$$P\{R\} = P\left\{ \frac{Y_i}{\sigma} \geq a_0 \right\} P\left\{ \frac{Y_i}{\sigma} \geq a_1 \right\} \ldots P\left\{ \frac{Y_i}{\sigma} \geq a_n \right\}$$

[28] This statement is a direct consequence of the independence of the variables $Y_i, Y_{i+1}, \ldots$

When this formula is applied to the Monroe case one gets

$$P\left\{\frac{s_i}{\sigma} \geq 1.79, \frac{s_{i+1}}{\sigma} \geq 1.42, \frac{s_{i+2}}{\sigma} \geq 0.77, \frac{s_{i+3}}{\sigma} \geq 0.51\right\} = 2 \times 10^{-4}$$

In words, there is one chance in 5000 that such a pattern will occur in a purely random series.[29] A similar calculation can be performed in the Diana case, but in order to be on firm ground one would first have to explain what produces the sharp increase which occurs in November and December.

Rating the quality of the data Before closing this section, we would like to emphasize (once again) that prior to caring about statistical treatment of the data it is important to assess the quality of the data. This step is often omitted by social scientists. As we already mentioned, in physics the quality of experimental data is guaranteed by the fact that the same experiment is performed by several groups. Even once a phenomenon is well known it is not uncommon that new experiments are performed in order to improve the accuracy of the measurement. In the social sciences the quasi-experiments done by one researcher are almost never repeated and checked by others. As an illustration, we will try to rate the quality of the data used in Phillips's paper (1974).

The criterion that the suicides must be publicized on the front page of the *New York Times* seems to define them unambiguously. However, this definition is not as clear-cut as it might seem at first sight. Consider for instance the death of Marilyn Monroe. The title of the article in the issue of August 6, 1962 reads: "Marilyn Monroe dead. Pills near. Official verdict delayed." There was no qualification of suicide in the first announcement. Yet this case was included in the sample. On the contrary, the death of Ernest Hemingway in July 1961 was *not* included in the sample in spite of the fact that it became known subsequently that it was indeed a suicide.[30] The title of the article on the front page of the *New York Times* of July 3, 1961 reads: "Hemingway dead of shotgun wounds. Wife says he was cleaning his weapon." As in the Monroe case there is no qualification of suicide in this first announcement. Thus, it is not obvious why the two events should be treated in different ways.

More generally, one can observe that 22 of the 33 cases in the sample produce only 1.7% of the total number of excess-suicides, whereas there are 6 cases which produce as much as 54% of the total. This suggests that the criterion based on the *New York Times* is too wide a net in the sense that it collects a lot of irrelevant

---

[29] One reason why such an impressive figure should not be taken too seriously is precisely that we are not sure that the suicide numbers constitute a *purely* random series. For instance, it cannot be excluded that the relaxation pattern is due to a series of correlated exogenous shocks which mimic a decreasing output.

[30] In contrast to the death of Marilyn Monroe, Hemingway's death was *not* followed by an excess number of suicides.

events among which, almost by chance, there are a few which are of greater relevance.

The main practical messages of this chapter can be summarized as follows.

(1) It is important to optimize the signal to noise ratio in the early phase of the design of the quasi-experiment.
(2) The extreme value technique can be especially useful when one wishes to assert the reality and order of magnitude of the phenomenon under consideration.
(3) Before embarking on statistical tests it is appropriate to check the reliability and accuracy of the data (for instance by comparing them to similar data or making internal consistency checks). The question of data reliability is the subject of the next chapter.
(4) Every time one has some a-priori knowledge about the phenomenon under study it makes sense to use pattern matching techniques.
(5) In order to estimate the likelihood that a fluctuation of amplitude $Y_t = s$ is due to the noise background (rather than to a genuine signal) one must compute the standard deviation $\sigma(Y)$ of the time series Y_t with the best accuracy possible, which means over a time interval where Y_t is stationary and which is the longest possible.
(6) By repeating the experiment several times in the course of time or by resorting to spatial disaggregation (i.e. observing the phenomenon in different regions of the country under consideration) it is possible to generate many realizations. If these realizations are not positively correlated (a negative correlation is not an obstacle but on the contrary an advantage) the averaging process often allows a substantial enhancement of the level of significance.

3.6 Conclusion

In this chapter we explained and illustrated several methods and techniques for improving the signal to noise ratio. As several of our illustrative examples concerned the Diana–Werther effect, it may appear somewhat surprising and frustrating that we did not propose a definite conclusion regarding the existence of this effect. We are in the same position as physicists who try to detect gravitational waves (see the previous chapter) in the sense that if we had several dozen "big" events such as the Monroe and Diana cases, it would be possible to draw a fairly clear conclusion. For the deaths of less known female celebrities, the effect is just too small to be detected. However, the evidence is fairly significant at the global level for a sample of 33 events. It is by accumulating an ever larger number of tests that we will be able to draw a more definite conclusion.

In this chapter we suggested that in physics (thanks to the collective validation procedure) the reality (or non-reality) of a phenomenon can usually be established fairly quickly. This may be true in 90% of cases, but there are also some cases in which it takes decades or even centuries before a definite conclusion can be reached. As an example, one may mention the use of a divination rod in the discovery of underground mines or springs. This effect has puzzled physicists for over three centuries. It is only in recent times, thanks to the use of highly sensitive magnetometers, that the problem has received a preliminary answer (e.g. see Chadwick and Jensen 1971, Rocard 1981). Furthermore, one should keep in mind that even once a question has for the most part been solved there may be objections which cannot be answered adequately. For instance, the fact that the Earth moves around the Sun was well accepted by most scientists in the early nineteenth century, but there was still the objection that no apparent displacements of the stars could be observed as a consequence of the Earth's motion. The fact that these displacements do indeed exist, but are too small to be measured except for the nearest stars, became completely clear only in 1838 when Friedrich Bessel was able to measure the displacement (the so-called stellar parallax) of the star 61 Cygni. Such examples teach us the great virtue of patience. In conclusion, the problem of the Diana–Werther effect will perhaps provide an exciting stimulus for coming generations of sociologists and econophysicists just as the measurement of the stellar parallax did in astronomy.

4

Equilibrium and metastable states

Equilibrium phenomena are simpler to analyze than time-dependent phenomena. The fact that non-linear time-dependent systems may display perplexing behaviors is illustrated in Fig. 4.1. It does not come as a surprise, therefore, that the equilibrium requirement is an essential assumption in statistical mechanics. For time-dependent systems the notions of statistical ensemble, temperature, entropy as well as several other fundamental concepts are no longer clearly defined and many standard results of statistical thermodynamics, such as the equipartition law, are no longer satisfied. In a general way, whereas equilibrium statistical mechanics provides a systematic and unified theoretical framework, non-equilibrium statistical mechanics is rather a collection of loosely connected techniques and equations.[1]

It is certainly important to know whether or not a social system can be considered as being in equilibrium. What criteria do we have in this respect? This is the first question that we consider. For a system to remain in equilibrium there must be an equilibrium restoring mechanism, therefore one is naturally led to examine these mechanisms. This is done in the first part of this chapter. In the second part we investigate metastable systems. From a common sense perspective a diamond would seem to be a perfect example of a system in equilibrium; the fact that diamonds are in fact metastable shows that common sense can be fairly misleading. By discussing several examples of metastable systems in physics and in chemistry we try to develop a qualitative understanding of metastability; in particular we emphasize the role of facilitator played by "seeds" or "precursors" of the new organization. In the closing section of the chapter we turn to social systems; we outline two historical cases which suggest the important role of these "seeds" in the process by which systems leave a metastable state in which they have been trapped.

---

[1] One can mention the Boltzmann, Liouville, Langevin, Fokker–Planck and Vlasov equations.

Gas: $1/r^6$ van der Waals attraction forces, strong short range repulsive force

Gravitating bodies: $1/r^2$ gravitational attraction, no repulsive force

Fig. 4.1 The influence of long range interaction on the evolution of a system of interacting bodies. The first scenario may correspond to a system of molecules whereas in the second each dot may represent a galaxy. Because microscopic phenomena are dominated by short range interactions we are more accustomed to the first scenario than to the second. As the systems are supposed to be isolated, both evolutions correspond to an increase of entropy. This example, which was suggested by Penrose (1989, p. 338), emphasizes the fact that long range interaction can induce evolutions which are at variance with the common sense notion of entropy. The second scenario is explored in the framework of a specific model by Chavanis *et al.* (2002).

4.1 Equilibrium restoring forces

To be defined properly the notion of equilibrium requires that one specifies the scale (in both time and space) on which one wishes to focus. On a time scale of 10^{-11} seconds and for distances of the order of one nanometer, even a dilute gas in equilibrium would not *appear* to be in equilibrium. This reminds us of the fact that the notion of equilibrium is a macroscopic notion. The real difficulty is that one does not know where exactly one should place the dividing line between micro- and macro-systems. This is even more true for social systems than for physical systems. Even if it is not a sufficient condition of equilibrium, the existence of a well-defined equilibrium restoring mechanism is certainly a necessary condition.

The rapid decay of extreme fluctuations in physical systems can be illustrated by the example of the relaxation process of neutrons in a nuclear reactor. In the process of fission, neutrons are released which have an energy of 2 MeV, which corresponds to a velocity of 177,000 km/s. These so-called fast neutrons will then undergo collisions with the atoms which surround them. In each of these collisions a neutron will lose a part of its energy to the atom with which it collides; this atom in turn will share its excess-energy with its neighbors. Thus, we have a cascade

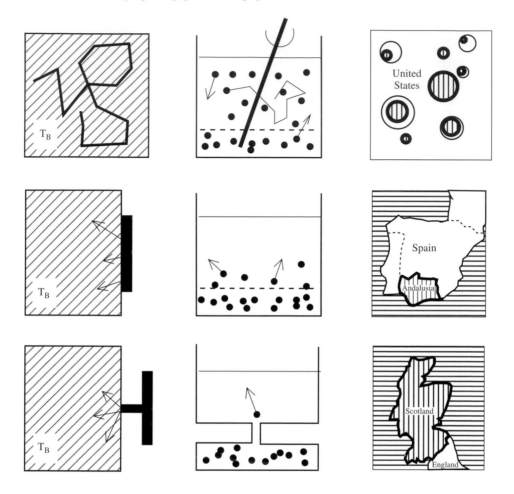

Fig. 4.2 Homogenization of temperatures, densities or linguistic differences. The first column describes the relaxation toward equilibrium of a small part (in black) of a system. The bulk of the system (hatched) acts as a heat bath. From top to bottom: (i) The thermalization of fast neutrons in a nuclear reactor is extremely swift because each neutron is able to interact with the surrounding atoms. (ii) The thermalization is much slower when the interaction occurs only in a thin layer of the surface of contact between the systems. (iii) When the surface of contact is reduced the relaxation toward equilibrium becomes even slower.

The figures in the second column show the same mechanism for unequal initial densities. The system can be thought of as consisting of a mixture of glycerol $(A, d_A = 1.26)$ and water $(B, d_B = 1.0)$. The two liquids are miscible in all proportions but the homogenization time is conditioned by contact conditions. From top to bottom: (i) The mixture is stirred with a stirring rod which plays the same role as the velocity of the neutrons, bringing the molecules together and enabling them to establish A–B bonds. (ii) When the mixing is left to the action of diffusion, the homogenization is much slower. (iii) When the surface of contact is reduced the relaxation toward equilibrium becomes even slower.

process through which the fast neutrons progressively share their energy with the surrounding atoms. This process is called thermalization because at equilibrium the velocity of the atoms and of the neutrons is fixed by the equipartition law $(1/2)m_a\overline{v_a^2} = (1/2)m_n\overline{v_n^2} = (3/2)kT$, where m_a, m_n and v_a, v_n are the masses and velocities of an atom and a neutron respectively, k is the Boltzmann constant and T the Kelvin temperature. For a temperature of 300 degrees Kelvin this condition gives the thermal neutrons an average velocity of the order of 2,000 m/s. Thus, in order to reach their equilibrium state the speed of the neutrons must be reduced by a factor: $170 \times 10^6/2,000 = 85,000$. How long does this process take? In water it takes 16 collisions, in graphite it takes 91 collisions. On the basis of a mean time between collisions of the order of 10^{-12} s the relaxation time to equilibrium is less than 10^{-10} s. This result can also be expressed in terms of temperature. Indeed, if we apply the equipartition law $E = (3/2)kT$ to the fast neutrons before their thermalization, we can see that an energy of 2 Mev corresponds to a temperature of 24×10^9 degrees Kelvin.

One may wonder why the relaxation time is so short in this case. A qualitative answer is provided by Fig. 4.2. The figure emphasizes that a crucial condition for thermalization to occur quickly is that the two populations of particles must be able to interact. In the fast neutron case, 100% of the neutrons are in direct contact with the cooler atoms. On the contrary, if droplets of molten iron are immersed into a tank of cold water, only the iron atoms which are located within the outer layers of the droplets will be in contact with the atoms of water. Thus, the proportion of interacting atoms will be $4\pi r^2\delta r/(4/3)\pi r^3 = 3\delta r/r$; if δr is of the order of several atom diameters, the ratio $\delta r/r$ will be very small for droplets of macroscopic size. Moreover, the proportion of interacting atoms decreases as r becomes larger, which is consistent with the observation that the smaller the

Caption for Fig. 4.2 (cont.) The figures in the third column illustrate the process of linguistic homogenization through which a population of immigrants A progressively drops its mother tongue and adopts the language spoken by the rest of the population. The driving force of the shift is the necessity for A people to speak the language B whenever they come into contact with B people in schools, jobs, hospitals or administrations. This is why the relaxation time is conditioned by contact opportunities. From top to bottom: At the end of the nineteenth century there were many German-speaking communities (represented by hatched thick lined circles) in US cities. In this case the relaxation time is about one generation. The (relative) rapidity of the process is due to the fact that the surface of contact is much larger than in the case of Andalusia with respect to Spain or Scotland with respect to England. It took centuries for English to diffuse into Scotland. These situations can be quantified by defining an index of spatial contact, which turns out to be equal to 1, 0.77 and 0.17 respectively (more details can be found in Roehner and Rahilly (2002) and in Alesina and Spolaore (2003)).

droplets, the faster the cooling. This rule is illustrated in the first column of Fig. 4.2.

The previous argument also applies to other transport phenomena such as diffusion, which leads to homogenization of densities or momentum transfer. The case of diffusion is illustrated in the second column of Fig. 4.2. A simple example of momentum transfer is the drag experienced by rain drops as they fall in the atmosphere; the drops lose momentum to the surrounding air through viscous stresses and decelerate. As one knows, when the droplets are small enough (e.g. droplets of mist) the water particles decelerate to the point of remaining suspended in the air. The third column in Fig. 4.2 shows a similar case in the social sciences. It illustrates the fact that the characteristic time of language diffusion crucially depends on the contacts between people speaking a language A and those speaking a language B. If the As do not live in closed communities, each individual will be permanently in contact with B people; in this case, which is similar to the fast neutron case, "thermalization" may take a few years. The upper panel in the third column illustrates a more realistic case where the A people form small communities which are immersed into B-speaking towns and cities. Observation shows that in such a case the "thermalization" takes of the order of 20 years. If the A people form a massive community, as in the cases of Andalusia and Scotland, the "thermalization" can take several centuries.[2] Needless to say, when the "thermalization" takes such a long time, its speed (and even its direction) can be modified by many macro-historical events.

In addition to the surface of contact, another important parameter is the strength and speed of propagation of the restoring forces. This is discussed in the next section.

4.2 Probing the strength of equilibrium restoring forces

In the context of economics, it is the mechanism of arbitrage which is the main equilibrium restoring force. Arbitrage is a very common practice in everyday life. Suppose that I wish to buy a television set. First I visit store A where it costs $p_A = 300$ euros, then I find another store B where it costs only $p_B = 280$ euros. Naturally, I will pick up the second one. This, however, is a gedanken-experiment, rather than a real one. For arbitrage to be defined precisely one must ensure that the two products are exactly identical. If, for the sake of argument, we assume that store B is located in a remote suburb whereas store A is located in the part of the city where I live, the two products will no longer be identical from

---

[2] In Roehner and Rahilly (2002, p. 109), the slow changes of linguistic borders are described and traced back to the sixteenth and fifteenth centuries.

my perspective. The cost of driving to store B should be added to the price tag p_B. Apart from transport costs there may be other differences between the two stores. One store may offer customers free entry into a parking lot or free home delivery of the TV set. In practice it is very difficult to identify and estimate all the differences. A simple procedure is to consider that the mean long-term price differential $\overline{p_A - p_B}$ somehow represents the "structural" price differential for the two products. In this interpretation the fluctuating component of the price differential can be written $\delta p = (p_A - p_B) - \overline{p_A - p_B}$. If one is only interested in the standard deviation or in the autocorrelation of δp, the constant term $\overline{p_A - p_B}$ will play no role and can be omitted.

For commodity markets, arbitrage works in much the same way as in the example of the TV sets. Figure 4.3 shows daily wheat price differentials for Houston (Texas) and Omaha (Nebraska). In this case the constant price differential $\overline{p_A - p_B}$ is equal to $473 - 410 = 63$ cents per bushel. It is of course not surprising that the price is lower in Nebraska than in a port like Houston from which the wheat can be easily exported. The wheat in Nebraska is similar to the TV set located in a remote suburb. In both cases the price is lower, in the same way as the price for any product is lower at factory level than in retail stores.

For arbitrage on financial markets one must keep in mind that most financial products have no intrinsic usefulness. Once bought, their only usage is to be sold

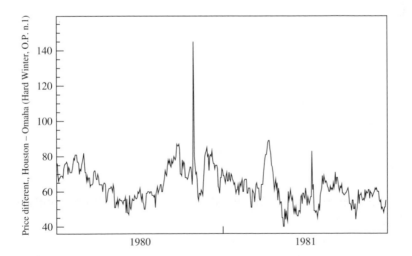

Fig. 4.3 Daily price differentials between two wheat markets. Vertical scale: price difference between Houston (Texas) and Omaha (Nebraska) expressed in cents per bushel. The average price differential is 63 cents, which basically corresponds to transport costs between the two markets. Hard Winter means that it is winter wheat, O.P. means Ordinary Protein. *Source: US Department of Agriculture (cash grain prices).*

or redeemed. This can be illustrated by the example of IBM shares on different markets. In 2006, IBM shares were traded on 14 stock exchanges in the United States and Europe. However, apart from the New York Stock Exchange where the main transactions took place, only three other markets were really active: Frankfurt, Stuttgart and the German electronic market Xetra.[3] Are the IBM shares traded in Frankfurt and in New York really identical? If they are they must produce the same dividends. This was not the case however. Over the period 2000–6, the NYSE shareholders received a dividend every quarter whereas the Frankfurt stockholders got dividends only until August 2001 (altogether they got four dividends). Furthermore, a Frankfurt IBM share cannot be directly exchanged against a NYSE share because the first is priced in euros while the second is priced in dollars. This is why only Frankfurt and Stuttgart are considered in Fig. 4.6. Nonetheless, it is well known that the prices of IBM shares in New York and Frankfurt move in parallel.[4] This can be attributed to two factors: (i) the fact that it is a quasi-arbitrage situation; (ii) the fact that the exogenous shocks that affect the two shares are certainly highly correlated.

If one considers the prices of shares of *different* stocks, various kinds of arbitrage mechanisms can be at work. As a matter of fact, the term arbitrage seems too narrow to provide an appropriate description in this case. Arbitrage is only one of the components in the strategies developed by investors. It is a complex game in which hedging techniques, risk reduction tactics and outpricing strategies play a great role.[5]

Figures 4.4 and 4.5 explain how it is possible to quantify equilibrium-restoring forces by using the autocorrelation of the price fluctuations. Figure 4.4 explains this technique on a simulation in which increasing amounts of noise were added to the relaxation pattern of a deterministic signal. The system is described by a first order stochastic recurrence equation whose deterministic part is $y_t = a y_{t-1}, 0 < a < 1$. For such a process the strength of the equilibrium restoring force is defined by $f(a) = (1 - a)/a$. In the limit $a = 1$ there is no restoring force and the process diverges linearly. However, when $a = 0.1$ (i.e. $f = 9$), $y_{t-1} = 10 \Rightarrow y_t = 0.1 \times 10 = 1$; in words, a deviation from equilibrium equal to 10 will at the following step be reduced to 1; when $a \longrightarrow 0, f \longrightarrow \infty$. Two conclusions emerge from Fig. 4.4. (i) The shape of the autocorrelation function is the same whether one is in a transient, non-stationary situation as in the first column or in a fairly stationary situation as in the third column. This is fairly uncommon in the sense that many statistical tools and techniques apply only to stationary

---

[3] On the London Stock Exchange there was only one transaction every 3 or 4 days. On the Paris Bourse (Euronext) the quotation of IBM was terminated on September 16, 2005.
[4] See in this respect Roehner (2000a, p. 177)
[5] Case studies of strategies developed by investors with respect to K-Mart and Converium shares can be found in Roehner (2005b)

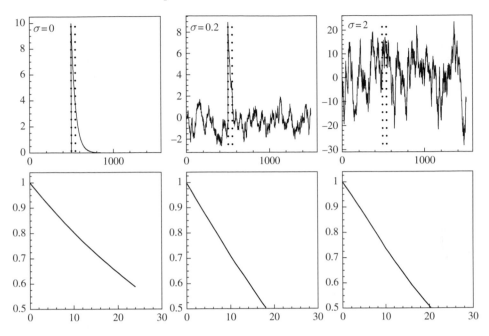

Fig. 4.4 Effect of an equilibrium restoring force in increasing levels of noise. The time series corresponds to a stochastic recurrence equation:

$$Y_t = aY_{t-1} + N_t + \delta_{t,t_1} \quad a = 0.98$$

where N is a Gaussian random variable of mean zero and standard deviation σ. The graphs in the second line show the autocorrelation function as a function of time lag. The graphs in the first column correspond to a non-stationary process whereas those in the third column correspond to a process which is fairly stationary. The present simulation as well as the analytical calculation show that the autocorrelation is independent of σ and given by $\rho(r) = a^r$, which corresponds to a relaxation time $\tau = 1/\ln(1/a)$. Intuitively this makes sense because the noise term can be considered as a combination of impulse-like shocks.

processes.[6] (ii) The relaxation time $\widehat{\tau}$, which is derived from the decay of the autocorrelation function $\rho(r) = \exp(-r/\widehat{\tau})$, indeed provides a reasonable estimate of the equilibrium restoring dynamics defined by the parameter a. Indeed, if the strength of the interaction $f = f(a)$ and $1/\widehat{\tau}$ are plotted on the same graph, the two curves are fairly parallel.[7] In other words, the parameter $1/\widehat{\tau}$ derived from the shape of the autocorrelation function provides a reliable estimate of the strength of the equilibrium restoring force. This interpretation is confirmed by the fact that for geographically separated wheat markets, there is a

[6] Naturally, we do not mean that this result holds for any kind of transient regime. Nonetheless, the pattern shown in Fig.4.4 may have a validity which extends beyond the assumptions made in the model.
[7] The analytical result for $1/\tau$ is $1/\tau = \ln(1/a)$ which is closely parallel to the curve of $f = f(a)$.

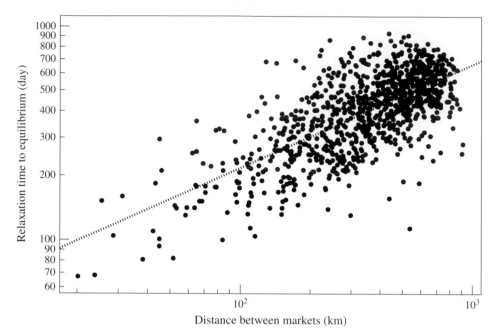

Fig. 4.5 Relationship between relaxation time and distance. Each data point corresponds to a pair of wheat markets in nineteenth century France. There are 45 markets (990 pairs); the period is 1825–49 with prices being recorded every two weeks. The correlation of the scatter plot is 0.69 (confidence interval is 0.65 to 0.72 for $p = 0.95$); the slope of the regression line is 0.48 ± 0.03, which gives the relationship: relaxation time $= A(\text{distance})^{0.48}$. The correlation between distance and relaxation time suggests that the latter is a good measure of the equilibrium restoring force because the smaller the distance between markets the easier it is for traders and wholesalers to arbitrage price differences *Source: Drame et al. (1991).*

strong intercorrelation between the relaxation time for pairs of markets and the distances between them (Fig. 4.5).

Figure 4.6 shows that the estimates of the equilibrium restoring mechanism given by the autocorrelation function can be used to describe a broad spectrum of situations. Incidentally, Fig. 4.6 is an illustration of the extreme value technique presented in an earlier chapter. It is because these situations are very different that τ takes values which differ by several orders of magnitude. This methodology for estimating interactions between markets complements the correlation length approach discussed in Chapter 1.

In the next sections we consider the case of metastable states, which play an important role in physics as well as in the social sciences.

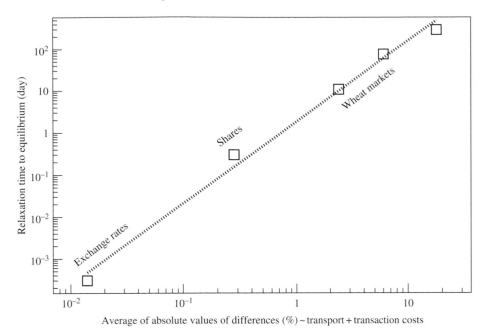

Fig. 4.6 Relationship between relaxation time and price differentials: The graph shows (from right to left) five sorts of arbitrage processes for transactions characterized by decreasing "viscosity". (i) Two distant wheat markets (Louviers–Toulouse, 1825–49, $d = 625$ km). (ii) Two closer wheat markets (Louviers–Rouen, 1825–49, $d = 30$ km). (iii) Two US wheat markets (Houston–Portland, 1980–1, $d = 3000$ km). (iv) Quotations of an IBM share on two German stock markets (Frankfurt–Stuttgart, 2005–6). (v) Dollar–euro exchange rates on two financial markets (London–Singapore, March 2006). Exchange rates are one of the most liquid financial products because dollars can be exchanged against euros on any market, whereas for shares the transaction requires (at least) three operations: sale of shares on market *A*, currency conversion, purchase of shares on market *B*. The slope of the regression line is 1.9. *Sources: French wheat prices: Drame et al. (1991); American wheat prices: US Department of Agriculture, Cash grain prices; IBM shares: http://finance.yahoo.com; exchange rates: http://www.forex.directory.net/quotesfx.html.*

4.3 Metastable states in physics and chemistry

To introduce the notion of metastable state we first describe the example of the solidification of water. The experiment can be conducted as described in Fig. 4.7. A small transparent plastic bottle is filled with water and put into the freezer compartment of a refrigerator at −18 degrees Celsius. After about 30 minutes it is taken out; with a little good luck the water is still liquid but is at a negative temperature. As a test one can either submit the bottle to a shock or, as shown in Fig. 4.7, add a small crystal of ice. Starting from this seed, ice will form and progressively fill the bottle. The speed of solidification depends upon the initial

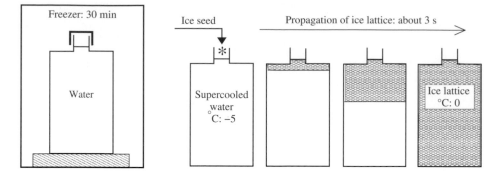

Fig. 4.7 Key role of "seeds" in bringing metastability to an end. If the ice crystal seed is replaced by another particle (e.g. a sand grain, a crystal of salt, a particle of wood or clay) metastability is not broken (except if the particle is too big, but in this case it is the shock which matters). This experiment suggests that the introduction of an ice crystal works because it provides a template for establishing the kind of bonds that exist in ice. The quantitative indications given in the figure refer to a small bottle of about 4 cm in height and 2 cm × 2 cm in section. The more negative the initial temperature $-\theta$ the greater the speed of propagation v of the solidification front; for instance $\theta = 1$ results in $v = 1$ cm/s, and $\theta = 5$ in $v = 3$ cm/s, more generally the rule is $v \sim \theta^{1.5}$. Incidentally, it can be noted that if one tries to record the "temperature" during the solidification phase, one is confronted by the fact that the measurement is device dependent; the indications of an alcohol thermometer, which has significant thermal inertia, will lag behind those given by a thermocouple thermometer, whose wires have much smaller inertia. This illustrates the fact that the very notion of temperature has no clear, univocal definition in time-dependent systems. More details can be found in Fujioka (1978). In the social sciences rapid transitions similar to the end of a metastable state have been described for instance in Sornette (2003) and Turchin (2003b).

temperature: the lower the temperature of the water the faster the process of solidification. Note that the ice which is formed is not "normal" ice. Due to its rapidity, the transformation is almost adiabatic; therefore the amount of heat generated by the transformation of water into ice (i.e. 80 calories per gram of water) must be absorbed by the increase of the temperature of water from its initial value, say -5 degrees, to 0 degrees (i.e. 1 calorie per gram and per degree); consequently the ice which is formed contains less than 10% "real" ice. As a confirmation, it can be observed that if left at room temperature the ice transforms into water in one or two minutes whereas it would take about ten minutes for such an amount of "real" ice to melt. Water which remains liquid at negative temperature is called supercooled water. Physicists have succeeded in supercooling water to about -40 degrees Celsius.

As a matter of fact, metastable states are very common in physics and in chemistry. The various allotropes of carbon constitute one of the most puzzling examples. As one knows, carbon exists in the form of graphite and diamond and it is graphite which is the stable form at room temperature and pressure.[8] Diamonds have a higher enthalpy ($\Delta H = 1.9\,\text{kJ/mole}$) and a higher free energy ($\Delta G = \Delta H - T\Delta S = 2.9\,\text{kJ/mole}$). Thus, according to the rules of thermodynamics diamonds should spontaneously transform into graphite. But transformation of this kind can be excessively slow. As another illustration one can mention the myth about the tin buttons on the uniforms of Napoleon's soldiers during the Russian campaign of 1812 (it can be read on several websites). The myth is based on the scientific fact that tin exists in two allotropic forms: white tin, which is a silvery metal used in soldering, and gray tin, which is in fact a metal powder. The transformation of white tin into gray tin is said to occur at temperatures below 13 degrees Celsius. According to the myth, a consequence of the low temperature of the Russian winter was that the tin buttons of the uniforms disintegrated into powder with the disastrous consequences that one can imagine. However, the fact that the transformation of white tin into gray tin is very slow can be seen in two ways. (i) Tin dishes that are left in unheated attics for several decades do not disintegrate into gray tin. The transformation only produces some spots where the metal is eroded but the indentations are less than a millimeter deep. This argument may perhaps not prove conclusive because tin dishes are in fact made of an alloy of tin, but then the tin buttons were probably also made of an alloy. (ii) Experiments conducted by physicists show that a test-cylinder of pure tin placed in a freezer at -18 degrees for 18 months is only slightly eroded. As in the case of the dishes, the indentations are less than one millimeter deep (Kariya *et al.* 2000).

Before coming to examples of metastable states in chemistry, one can mention a case at the border between physics and chemistry. It is the supersaturated solution of sodium acetate. It is because the solubility of the compound decreases with temperature that a saturated solution becomes supersaturated when it cools down. The case is at the border of physics and chemistry because sodium acetate molecules form clusters containing three molecules of water, the so-called molecules of sodium acetate trihydrate.

In principle, thermodynamics tells us whether or not a chemical reaction will occur. However, in many cases a reaction which should occur does not. The following list gives a few simple examples.

- A mixture of hydrogen and oxygen at room temperature is perfectly stable. Thermodynamically, however, the mixture is highly unstable in the sense that its free energy is much higher than that of water:

$$G\,(2H_2 + O_2) - G\,(2H_2O) = 237\,\text{kJ/mole}$$

[8] Fullerene molecules, an another metastable form, were discovered in 1985.

The same results applies to a mixture of hydrogen and chlorine:

$$G\,(H_2 + Cl_2) - G\,(2HCl) = 92\,kJ/mole$$

- One might think that because decomposition is an "easier" process than recombination, it is less likely to give rise to metastable states. However natural it may be, the idea is refuted by observation. One can mention the two following examples. (i) Hydrogen peroxide, H_2O_2, has a shelf-life of several weeks in spite of the fact that the reaction $2H_2O_2 \longrightarrow 2H_2O + O_2$ is highly exothermic. (ii) The reaction $2NO \longrightarrow N_2 + O_2$ should occur spontaneously because

$$G\,(2NO) - G\,(N_2 + O_2) = 175\,kJ/mole$$

yet nitric oxide can be stored indefinitely at room temperature and pressure without detectable decomposition. This example is even more striking than the previous ones because NO as well as N_2 and O_2 is a gas at room temperature. The transition from one gas to another may seem an "easier" step than the transformation of gases into liquids or of liquids into solids.

When a system is thermodynamically unstable but does not undergo any transformation it is said to be kinetically stable, which is just another way to say that the transformation does not take place.[9] Metastability can also be expressed in terms of the activation energy E_a, which is defined by the relation

$$K = A \exp(-E_a/kT) \tag{4.1}$$

where K denotes the rate constant of a reaction. When the rate constant is known for different temperatures, the activation energy can be estimated from Equation (4.1); however, as far as we know, activation energies cannot be computed theoretically from first principles.

Usually, metastability can be broken fairly easily.[10] The following list summarizes some means through which this can be done.

- Supersaturated solution of sodium acetate $\longrightarrow$ crystallized form of sodium acetate: add a small crystal of sodium acetate.
- Water $\longrightarrow$ ice: add a small crystal of ice, provoke a shock.
- $2H_2 + O_2 \longrightarrow 2H_2O$: add palladium powder or provoke a spark.
- $H_2 + Cl_2 \longrightarrow 2HCl$: put in ultraviolet light.
- $2H_2O_2 \longrightarrow 2H_2O + O_2$: add manganese oxide powder.

Can we find some broad qualitative rules which may suggest possible clues for the social sciences? The solidification of water is a process which is not very

---

[9] In chemistry the kinetics of a reaction designates its speed, thus the expression "kinetically stable" means that the reaction is very slow.

[10] However the transitions from one allotropic form to another, e.g. diamond $\longrightarrow$ graphite or white tin $\longrightarrow$ gray tin, cannot be easily triggered.

different from the transition between allotropic states; indeed, at temperatures near the freezing point, molecules of water assemble in tetrahedron structures, whereas in ice these tetrahedrons are linked together to form an infinite lattice structure. To find its way from the fairly loose structure of water to the more complex structure of ice it is conceivable that the system needs some "guidance". The small crystal of ice which starts the solidification seems to provide that "guidance". Similarly, in its crystallized form, sodium acetate has a fairly complex lattice structure in which sodium acetate molecules are interspaced with water molecules. Again, it is conceivable that a small crystal of sodium acetate can provide a seed which facilitates the transition from the loose structure of a liquid to the more ordered structure of a crystal. A parallel to these crystal seeds is the role played by so-called chaperone molecules in the folding of proteins. Does this qualitative interpretation provide a better insight? At least, it helps us to understand why the transition from diamond to graphite is problematic. First, these are solids and solids are of course less flexible than liquids or gases; second, the two solids have very different structures, therefore a graphite seed would not provide enough guidance for the transformation of the diamond structure into the graphite structure.

In the next section we examine if these notions can help us to better understand some rapid transformations of social structures.

4.4 Metastability, seeds and forms of post-revolution societies

In physics and chemistry metastable states could be identified thanks to the fact that their energy is higher than the energy of equilibrium states. For social phenomena, as there is no similar means of identification, we must rely primarily on observation. The previous discussion suggests three possible criteria. (i) A metastable state is a situation where the system is frozen in a structure which seems outdated and at variance with the evolution of otherwise similar systems. (ii) The transition from the old structure to the new one is fairly rapid; this naturally suggests a parallel with revolutions. (iii) The presence of a "seed" helps to end metastability and at the same time provides a prototype on which the new society can be modeled. Do these rules provide a framework which may give us a better insight into some historical episodes?

Tsarist Russia A first example is tsarist Russia. Let us discuss the three previous points in succession.

(i) For the purpose of reconquering the Volga basin from the Tatars, Ivan IV (who reigned from 1547 to 1584) had to seek the help of the Russian aristocracy, the so-called boyars. But the expansion did not stop with the defeat of the Tatars;

it continued toward the south, south-east and east. This process was fairly similar to the process of the *Reconquista* which took place in Spain and it led to the same results: a strongly centralized state, an efficient standing army, and a rich and powerful aristocracy. The officers were usually rewarded by being granted huge estates either in the newly conquered territories or in the homeland itself. In contrast to Spain, where the conquest phase came to an end in the late sixteenth century with the conquest of the Philippines, in Russia it lasted until the end of the nineteenth century. One may remember that the last phase of this expansion was marked by recurrent wars against the Ottoman empire and by the Crimean war, which put an end to the expansion. This feature helps to understand why the autocratic regime and the privileged position of the aristocracy lasted much longer than in other European countries. In the second half of the nineteenth century, as economic expansion replaced geographical expansion, the entrenched privileges of the aristocracy became an obstacle to this transformation. But it is precisely because the previous phase had lasted much longer than in Western states that the power of the aristocracy was so deeply rooted. More than a century after the French Revolution, there was still no transformation in sight in Russia. The autocratic regime came to appear obsolete not only to Westerners but also to many members of the Russian elite. The rising of the Decembrists in December 1825 was the first in a long series of plots, assassinations and aborted insurrections. The insurrection of 1905 was a prefiguration of 1917 which almost succeeded.

(ii) What "seeds" were present in Russia by 1917? It is important to realize that the Russian reform movement of the late nineteenth and early twentieth centuries was not very different from those existing in Western countries. On the far left the Industrial Workers of the World (IWW) were present in many countries and especially in the United States. There were revolutionary socialists in Germany as well as in France and strong socialist parties existed in almost all Western countries. From this perspective the Bolshevik and Menshevik parties were by no means atypical. But the fates of these parties were very different in the West on the one hand and in Russia on the other hand. In France the far left suffered a great blow with the defeat of the Paris Commune; many radicals were killed in the battle for the control of Paris and an even greater number were deported after their defeat. In the United States the IWW unions were nearly suppressed through recurrent police raids followed by arrests and trials of the leaders. In Germany, the revolutionary socialists who had formed the Spartacus league were physically eliminated by the troops of the Freikorps in the aftermath of the Berlin uprising of January 1919; hundreds of members of the German communist party were killed. As we know, there was another major repression against the communist and socialist parties in 1933. On the contrary, in Russia the revolutionary socialists (i.e. the Bolsheviks) experienced a tremendous expansion. The political vacuum created by the abdication of the tsar and the social chaos brought about by the war provided ideal conditions for the growth of any organization able to fill the vacuum. To wage its war against an autocratic regime the Bolshevik party adopted a strong centralization and an organization modeled on the military. Thus, paradoxically, its characteristics were in line with those of the tsarist regime. This is probably why it was able to fill the vacuum so quickly.[11] As in the solidification of water, the seeds had a structure which was a compromise between the old and the new organization. Thus the party created by Lenin was particularly well suited to provide a firm anchor to this disorganized society. This form of organization was well adapted to the civil war of 1917–20; it was also a big asset in the tragic times which followed the German invasion of 1941. But although it proved efficient in times of war, this organization was not well suited for times of peace. Ironically, or perhaps significantly, it collapsed just one year after the end of the war in Afghanistan.

Naturally, this reconstruction omits many details and it can be argued that those which it includes were selected to suit the argument. It is true that to be convincing the "seed theory" should be able to explain other episodes apart from the case of Russia. This would require an extensive study which would lead us too far away from the topic of this book. Yet there are two additional points that we wish to make. First, we would like to respond to a possible objection. Second, we propose a prediction based on the seed approach. While reinterpreting the past is an exercise that can be done in a countless number of ways, trying to predict the future provides a test of greater significance.

The objection results from a comparison between the American and the Russian Revolutions. We explained that the centralized and authoritarian organization of the Bolshevik party was well adapted to wage the civil war against the anti-Bolsheviks. But one cannot forget that from Lexington (April 1775) to Yorktown (October 1781) the American War of Independence lasted over six years. Thus, if the seed argument is correct, how can one understand that the American polity which emerged from the War of Independence was so different from the Soviet polity which emerged from the Revolution? The answer comprises two parts. First, the structures of the two societies were very different before their respective revolutions. The Thirteen Colonies had almost no army. This can explain why the War of Independence was so long in spite of the fact that the British expeditionary force was inconsiderable, but it also explains why the nobility had little power in 1774. Had the War of Independence been more severe, it would have given an opportunity to army commanders to gain wealth and power. Although long, it was in fact a low-intensity conflict: there were less than 300 fatalities per month

---

[11] In contrast, the forces which brought about the Chinese Revolution of 1910 were altogether unable to fill the vacuum created by the end of the empire with the result that a long period of chaos followed.

(including the deaths due to illness). It can be estimated that the civil war in Russia had an intensity which was about 50 times greater than the American War of Independence. Thus, the American polity was only slightly modified by the war and retained most of the features that it had before 1775. It was at the federal level that a new structure had to be invented and it is therefore at this level that the seed theory may apply. The extensive powers granted to the President, and the fact that he cannot be dismissed by Congress (except by an impeachment procedure), are probably a legacy of the War of Independence. Moreover, the fact that in time of war no dissent and no dissenters should be tolerated became a permanent feature of America in times of war. The control of newspapers, the imposition of loyalty oaths, and the internment of enemy aliens became recurrent features, as could be seen during the Civil War,[12] World War I, World War II, the Cold War, and more recently in the period after September 11, 2001.

Saudi Arabia Finally, we close this chapter by examining a situation which at the time of writing (March 2006) appears to be metastable. As we know, there are no representative institutions in Saudi Arabia and in most of the Gulf States. In the absence of an elected parliament the political power rests entirely in the hands of the royal family. Moreover, the country comprises three groups between which there are only few interactions: (i) the royal family, which comprises some 10,000 members, and the ulemas of Wahhabism (also called Salafism); (ii) the Saudi citizens, many of whom are wealthy people but who do not have much say in political choices; (iii) the immigrants, many of whom have been in the country for decades yet cannot hope to become citizens because the economic privileges attached to citizenship cannot be extended to the whole population. These divisions are reminiscent of the situation which existed in France before the Revolution of 1789. At that time there were three groups: (i) the royal family, the aristocracy and the bishops who held the political power; (ii) the nobility who enjoyed special privileges in terms of tax exemption and access to military careers; (iii) the rest of the population which constituted the so-called Third Estate. In short, it was a segmented society. Such situations are potentially unstable because removing separations and barriers would lead to a society with more interactions. In physics such systems are characterized by a lower energy. Similarly, historical observation suggests that in their evolution social systems also try to maximize the number of their internal interactions.

If one agrees that the present situation is metastable, does this imply that a revolution will occur in the near future? Certainly not, for at least two reasons. (i) As long as oil revenue is abundant there will be no risk of the kind of bankruptcy

---

[12] About the suppression of dissent in the Civil War see for instance Adams (2000), DiLorenzo (2002), Goodrich (1995)

which so often marks the prefiguration of a political collapse. (ii) The struggle for power in Saudi Arabia is a matter of great concern for many countries and especially for the United States. Does the United States have the capacity to substantially influence the outcome of this political struggle? It is difficult to know because it is not in the interest, either of the United States or of the Saudis, to make this kind of information public. This opacity is in fact a great obstacle to an understanding of such issues. This is why we devote a subsequent chapter to the question of exogenous influence between states.

Is it possible to make more specific predictions? Both the French and the Russian models suggest that the years before the revolution are marked by a high level of political turmoil. In France it took the form of a struggle of the Parlement of Paris against the authority of the king for the preservation of its independence and privileges; in Russia it took the form of protests raised by the Duma and of clandestine action by groups of nihilists. What form can it take in Saudi Arabia? Clandestine action is the most likely, for there are few (if any) outlets for other forms of opposition. The struggle between Osama bin Laden and the royal family is one example of the form this opposition can take. If we are right in our diagnostic that this is a metastable situation, one would expect more actions of this type to occur in the future.

5

Are the data reliable?

It would be a waste of time and energy to analyze and try to make sense of flawed or unreliable data. Ensuring the reliability of experimental data is a cornerstone of the natural sciences. In physics there is a proven procedure through which the validity of experimental results is established. Every time researchers claim to have observed a new physical phenomenon other teams around the world try to replicate the results. If their observations are consistent with those obtained by the first team, the phenomenon will be given the status of a new physical effect, often named after its discoverers. The Foucault pendulum (1851), the Zeeman effect (1902), and the Aharonov–Bohm effect (1959) are a few examples among many. On the contrary, if the claimed result cannot be replicated, the physics community comes to the conclusion that the first observation was spurious. A case in point was the flawed discovery of cold fusion in 1989, which will be described in some detail below. In what follows, this phase of verification will be referred to as the *replication procedure*. It provides an efficient way through which flawed observations can be eliminated. Unfortunately, there is no similar procedure in the social sciences even in cases where replication would be possible. In a previous chapter we mentioned the fact that thirty years after its discovery the reality of the Werther effect is still not clearly established. This study could have been replicated in all industrialized countries in which monthly suicide statistics are published. It is true that related studies had been done, but most of them had little or no overlap with Phillips's seminal paper (1974). In short, the efforts of researchers were juxtaposed and dispersed rather than being focused on a definite issue. Accumulating data until the evidence becomes overwhelming does not seem to be a top priority on the agenda of social scientists.

The chapter is organized as follows. First, we examine two replication episodes: one in which the initial observation was confirmed and another in which replications led to observations at variance with the first experiment. These episodes also emphasize several important features of the replication process, in particular the fact that the replications are by no means identical to the first experiment. The resemblance is restricted to a set of crucial parameters. Distinguishing between the parameters which are crucial and those which are not is precisely one of the main purposes and achievements of the replication phase. In the second part of the chapter we discuss two cases in which facts and data have been misrepresented. The first case describes a statistical glitch which resulted in errors of as much as 30% in the suicide rates for New York State. The second case-study concerns an historical issue, namely the relations between Japanese people and occupation forces in the wake of World War II.

5.1 The replication process for the Foucault pendulum experiment

Named after the physicist Léon Foucault (1819–68), the Foucault pendulum experiment was conceived to demonstrate the rotation of the Earth on its axis. Its principle can best be explained by assuming for the sake of simplicity that the pendulum is located at the North Pole. Will its plane of oscillation rotate with the Earth or will it remain pointing in the same direction while the Earth rotates underneath? The observation shows that it is the second alternative which is correct. This means that an Earth-bound observer will see the plane of oscillation of the pendulum describe a full circle in 24 hours. In other words the period of the phenomenon is $T = 24$ hours. Applied to a pendulum set up somewhere on the Equator, the same argument says that an observer will see no changes in the orientation of the plane of oscillation, which means that in this case T is infinite (Fig. 5.1). An interpolation argument tells us that for a pendulum set up somewhere between the North Pole and the Equator $T > 24$ hours. For instance, in Paris $T \simeq 32$ hours. The experiment was publicly performed by Foucault on March 31, 1851. It attracted great attention and in following months its replication was attempted in more than 10 European cities as well as in some non-European countries. Table 5.1 presents a sample of these experiments. Some failed, but many succeeded and within a few months it became clear that the phenomenon observed by Foucault was indeed real. This, however, did not stop the flow of experiments. As in all physical experiments, once the effect had been recognized as genuine the challenge shifted to improving the accuracy of the measurement. Table 5.1 shows that in the course of three decades the accuracy was improved by a factor of 10. This quest of ever greater accuracy is not restricted to only a few famed experiments, it really concerns all experiments, even the most unpretentious. As an illustration, one can mention the fact that the boiling temperature of propanol was measured no less than 140 times; the first result, published in 1864, gave $t_b = 90 \pm 4$ degrees Celsius, while the most recent was published in

Table 5.1 *Replication of the Foucault pendulum experiment*

	Month	Year	Location	Physicist	Length (m)	Mass (kg)	Accuracy (%)
1	Jan	1851	Paris	Foucault	2	5	
2	3 Feb	1851	Paris	Foucault	11	5	
3	31 Mar	1851	Paris	Foucault	67	28	7
4	Spring	1851	Paris	Dufour	20	12	7
5	May	1851	Bristol, UK	Bunt	16	24	3
6	Jun	1851	York, UK	Phillips	16		1
7	Oct	1851	Rio, Brazil	D'Oliveira	4	10	87
8		1879	Groningen	Onnes	1	5	0.5
9		1937	Benares, India	Dasannacharya	1		1.3

Notes: The table presents only a sample of the experiments which were performed. In 1851 alone the experiment was performed in a dozen European cities (Barsics *et al.* 1993, p. 3). With the exception of experiment 7 there was a steady improvement in the accuracy of the measurement. Some of the accounts contain perplexing statements; for instance D'Oliveira claims that he found a direction of oscillation in the south-west sector for which the plane of oscillation remains invariant.

Sources: 1: Foucault (1851); 2: Barsics *et al.* (1993); 3: Foucault (1878), 4: Dufour *et al.* (1851); 5: Bunt (1851); 6: Phillips (1851); 7: D'Oliveira (1851); 8: Onnes (1879), Schulz-Dubois (1970); Dasannacharya and Heymadi (1937).

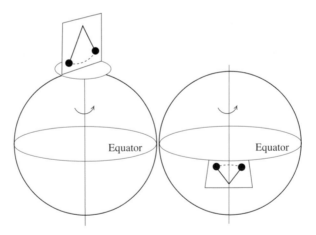

Fig. 5.1 Schematic representations of two extreme locations of the Foucault pendulum. The left and right figures represent a Foucault pendulum at the North Pole and on the Equator respectively. Observation shows that in the first case the oscillation plane describes a full circle in 24 hours whereas it remains fixed in the second case. This result is usually described by saying that the plane of oscillation remains pointing in the same direction with respect to the fixed stars. A question of interest is whether the observation would be the same if the Earth were spinning in an empty universe.

1997 and gave $t_b = 97 \pm 0.2$ degrees Celsius; in other words, over these 133 years the accuracy was improved by a factor of 20.[1]

Another point of interest is the fact that all the pendulums mentioned in Table 5.1 were in fact different. They differed in their mass, their length and in many other characteristics which are not mentioned in the table. For instance, experiments 1–3 used a steel wire suspension. In experiment number 6 Phillips first tried an untwisted silk suspension which had to be abandoned because it led to growing ellipticity (see below); Phillips then shifted to a catgut suspension which proved much more satisfactory.[2]

In conclusion, it must be emphasized that the replication procedure is by no means merely a repetition but is a creative process. It identifies the parameters that are essential and it covers a broad range of experimental conditions which were not considered in the first experiment. For instance, one may wonder if the effect is modified when the initial conditions are changed. The physicists in experiment number 4 found a difference of 13% depending on whether the initial plane of oscillation was in the north–south or east–west direction. Subsequently, however, this observation was dismissed.

There is a point which deserves particular attention in connection with the social sciences. In a general way, most physical observations involve more than one phenomenon. The Foucault pendulum is no exception. In fact the effect due to the rotation of the Earth is in competition with another phenomenon called the Puiseux effect.[3] The Puiseux effect says that when the trajectory of a pendulum can be approximated by an ellipse, the major axis of the ellipse rotates with an angular velocity given by the formula $\omega_P = \left(3\sqrt{g}/(8\pi)\right)(S/L^{5/2})$, where S is the area of the ellipse, L the length of the pendulum, g the acceleration of gravity. This effect has no connection whatsoever with the rotation of the Earth. Taking it into account, one finds that at the latitude of Paris, with $L = 67$ meters, and with a major semi-axis $a = 3$ meters, the angular velocity of the plane of oscillation ω is given (in 10^{-5} radians) by the expression $\omega = 5.5 \pm 0.2b$ where b is the minor semi-axis of the ellipse expressed in centimeters. Thus, as soon as $b > 5\,\mathrm{cm}$, the effect due to the rotation of the Earth is substantially altered and when $b > 27\,\mathrm{cm}$ one may observe a rotation of the plane of oscillation which is counter-clockwise instead of being clockwise. Many causes, such as asymmetries of the pendulum, vibrations of the point of suspension, draft and air turbulence, contribute to increase b, which is what makes the experiment fairly tricky. For Foucault and his followers the challenge was to isolate the effect due to the

---

[1] All these measurement data along with the references of the corresponding papers are available online at http://webbook.nist.gov.

[2] Catgut is a strong cord made from the dried intestines of sheep; it is used in particular for the strings of musical instruments.

[3] Victor Puiseux, French mathematician and physicist, 1820–83.

Earth. Similarly in the social sciences the real challenge is often to isolate the phenomenon that one wishes to study from other competing effects.

5.2 The replication process for cold nuclear fusion

For over 50 years, physicists have been trying to achieve controlled fusion. In contrast to nuclear fission in which one big nucleus is broken into smaller parts in a reaction which releases energy, in fusion two small nuclei merge thereby releasing a huge amount of energy. One possible reaction is $D + D \longrightarrow He + 24\,MeV$ where D represents a nucleus of deuterium composed of one proton and one neutron and He is a nucleus of helium which has two neutrons and two protons. So far, most attempts have tried to achieve hot fusion in which the temperature reaches several million degrees. In March 1989, two chemists at the Universities of Utah and Southampton held a press conference where they reported that electrolysis of heavy water had led to production of excess heat that could only be explained by a process of nuclear fusion. Naturally, the announcement aroused considerable interest. However, from the very beginning it was obvious that the reaction (if any) occurred only under special circumstances. Indeed, the electrolysis was kept functioning for several weeks and most of the time nothing noticeable was to be observed. It was only very occasionally that the temperature suddenly rose to about 50 degrees Celsius, remaining at that level for a couple of days.

Immediately after the announcement, teams of scientists around the world started work to replicate the experiment. For the next six weeks claims and counterclaims kept the topic on the front pages of the newspapers. By the end of May, as there was a growing body of failed replications the initial enthusiasm began to fade in the physics community. The title of an article which appeared on the first page of the *New York Times* on May 30 gives an indication of the climate at that time: "At conference on cold fusion, the verdict is negative." As of 2005, that is to say 16 years later, the situation is very much the same as it was by the end of 1989: cold fusion remains elusive but there is still no satisfactory explanation for the bursts of heat that have been observed episodically. Physicists and chemists in several countries continue the research and meet at the annual International Conference on Cold Fusion. On December 2, 2004 the US Department of Energy published a report which says that evidence on cold fusion remains inconclusive.[4]

Because of its negative outcome this episode seems at first sight to differ greatly from the previous one. However, if one discards the question of the outcome, the patterns of the two replication processes appear to be very similar. In both cases (i) there was great initial excitement among physicists; (ii) the first experiments

---

[4] New cold fusion claims were made in March 2004 which were based on compression induced by ultrasonic vibrations. The technique is called sonofusion.

brought mixed results; (iii) the question was basically settled within a few months; (iv) research on the topic continued albeit at a slower pace during the following decades. This second episode also has a number of specific features which deserve to be mentioned.

- Whereas the Foucault experiment was consistent with theory, cold fusion, in contrast, was at variance with theory. Nevertheless, researchers preferred to trust new experimental results. This reaction is based on the conviction that no matter how well established a theory may be, some unknown physical phenomena may be observed which will upset that theory.
- When the first duplication attempts failed, the first reaction of the researchers was to think that some necessary details had been omitted in the published paper. In other words, the researchers suspected that their failed experiments were not really replications but different experiments. It is not impossible that there was indeed a hidden factor which had escaped the attention of the "discoverers" themselves. Even small details can be of crucial importance.
- The fact that despite a steady stream of negative results this topic continues to attract the attention of scientists attests the persistence and patience of physicists and chemists. The situation is very different in the social sciences where even promising fields are abandoned in favor of newer, possibly more fashionable topics. How can one explain such different attitudes? The feeling shared by most physicists that perseverance is essential and in the long run rewarding is certainly a key factor. Perseverance is needed to gain a thorough understanding. Even if initially confined to a rather narrow topic, this understanding usually brings about broader progress. Wien's law initially had rather limited scope but eventually it opened the way to the development of quantum physics. On the contrary, any wrong yet unquestioned observation has the opposite effect: instead of opening new horizons it generates confusion and false starts. This is why it is essential, especially in the context of the social sciences, to weed out misrepresentations and faulty data.

The next two sections provide two examples of situations where faulty evidence can mislead researchers.

5.3 Biased suicide statistics

In this section we examine suicide rates in New York State and New York City in the 1980s. There are several possible sources: (i) the New York City Department of Health; (ii) the Centers for Disease Control (CDC); (iii) the yearly volumes of the *Statistical Abstract of the United States*. One would expect the data given by these sources to be in agreement. As shown in Fig. 5.2 there are substantial differences: 17% in 1984, 29% in 1985, 26% in 1986. This raises two questions.

- Which one of the series is correct?
- How can one explain the discrepancies?

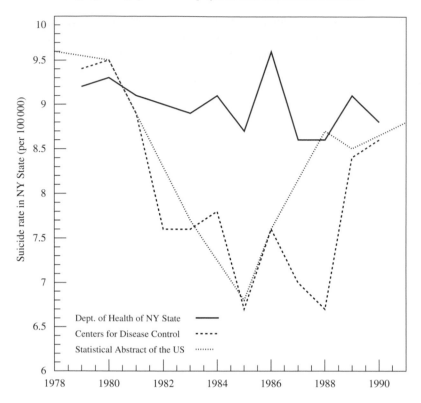

Fig. 5.2 Suicide rates in New York State. The graph shows annual suicide rates in New York State according to three different sources: the Department of Health of New York State, the Centers for Disease Control and the *Statistical Abstract of the United States*. The discrepancy between the first source and the two others reached 29% in 1985. *Sources: New York State Department of Health, CDC-WONDER database, Statistical Abstract of the United States (various years). Many thanks to Dr. Peter Herzfeld of the New York State Department of Health for sending me the data for New York State.*

The first clue came from the observation that for three of the boroughs which compose New York City there are huge differences between the statistics of the New York City Department of Health and the statistics provided by the CDC. As shown in Table 5.2 the differences are of the order of 100%. When one remembers that taken together these counties have a population which represents 28% of the population of New York State, it becomes clear that these discrepancies can explain the differences observed in Fig. 5.2. This represents a first progress in the sense that we have been able to narrow down the question to 3 counties instead of the 62 which compose New York State. But we still do not know the reasons for the differences.

Table 5.2 *Number of suicides in the boroughs of Brooklyn, Bronx and Manhattan*

	1980	1981	1982	1983	1984	1985	1986	1987	1988
Brooklyn									
NYC DH	149	137	141	142	111	134	107	127	124
CDC	156	145	140	139	91	47	33	56	73
Bronx									
NYC DH	107	97	68	87	80	82	96	67	90
CDC	112	102	44	35	34	29	30	14	22
Manhattan									
NYC DH	213	175	184	154	171	140	161	136	146
CDC	234	201	133	68	77	54	46	32	44

Notes: The table compares the number of suicides given by two different sources: NYC DH, New York City Department of Health; CDC, Centers for Disease Control. The borough of Brooklyn coincides with the county of Kings and the borough of Manhattan coincides with the county of New York. The trough (centered on 1985) displayed by the CDC data is a spurious effect which is explained in the text.
Sources: New York City Department of Health, CDC-WONDER database. I am grateful to Dr. Wen Hui Li, Director at the New York City Department of Health and Mental Hygiene for the data and additional explanations.

A further clue is provided by Fig. 5.3. In the International Classification of Causes of Death (ICD-9) suicide corresponds to the code numbers 950–959. Each of the intermediate numbers corresponds to a specific method of suicide such as hanging, poisoning, firearms, etc. The broken line corresponds to the code number 799.9 or in words "Other unknown and unspecified causes." Usually, this is a category which is not of much interest. Yet, Fig. 5.3 shows that there is a clear negative correlation between 950–959 and 799.9 (the coefficient of correlation is −0.68). Now we begin to understand what happened. All suicide deaths require medical determination which, quite understandably, may take some time, especially if detailed laboratory tests are required. In the meanwhile the deaths are registered as pending cases in the category 799.9. If for some reason, for instance a lack of qualified personnel, the determination process takes an unusually long time, it may happen that the provisional suicide numbers (not including the cases left in 799.9) are communicated to the CDC where they are put in the database. Once the new, revised suicide numbers become available at the level of the New York City Department of Health these new figures must also be entered into the CDC database to replace the old ones. For some reason, in the present case this was not done. Even more surprising is the fact that these faulty data also occur in the *Statistical Abstract of the United States.*

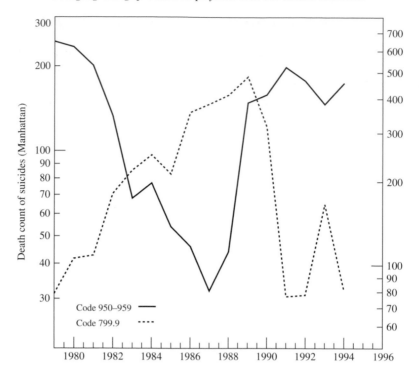

Fig. 5.3 Death counts in Manhattan for two codes of the International Class-ification of Death. Solid line: death count for code number ICD-9, 950–959: suicide, scale on left-hand side. Broken line: death count for code number ICD-9, 799.9: other unknown and unspecified causes of death, scale on right-hand side. This category was used by the New York City Department of Health in order to store pending cases until they had been treated. For some reason, the provisional figures became definitive data at the national level. The correlation between the two series is −0.68. From the difference in magnitude of the death counts it is clear that, apart from pending suicide cases, 799.9 also incorporates other pending cases such as murders or accidental deaths. *Source: CDC-WONDER (http://wonder.cdc.gov/mortlCD9).*

What conclusions can we draw from this case-study? It calls for increased vigilance in the handling of data. It would be a waste of time to attempt to interpret the huge trough displayed by CDC data for New York State (Fig. 5.4); as we have seen this trough is completely spurious. American statistics are probably the best in the world in terms of comprehensiveness and convenience of use. Yet in this case there has been a glitch. How can one check suicide data? As seen above a negative correlation between suicide and a vague "catch-all" category such as 799.9 should raise suspicion. Unfortunately, 799.9 is not the only category which may possibly contain ill-defined causes of death. Table 5.3 summarizes the situation in the old and in the new classification. In this respect it should be noted that even in recent years the death count of R99 showed huge fluctuations in the

Table 5.3 *Classifications of causes of death*

Classification	Unknown or unspecified cause of death	Suicide	Undetermined intent even after investigation
ICD-9	799.9	950–959	980–989
ICD-10	R99	X60–X84	Y10–Y34

Notes: ICD-9 and ICD-10 are the former (until 1998) and present International Classifications of Causes of Death. At the New York Department of Health the codes 799.9 and R99 were used to temporarily shelve cases which were pending investigation. The codes 980–989 (or Y10-Y34) are other categories in which pending cases can be stored. Any substantial increase in these "catch-all" categories may signal anomalies in the data collecting procedure and raise suspicion about data in categories such as "suicide" or "death through assault".

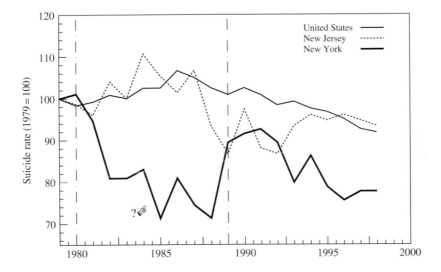

Fig. 5.4 Suicide rate in New York State, New Jersey and the United States. At first sight one may wonder what caused the 30% trough in New York State with respect to nearby New Jersey or to the rest of the United States As a matter of fact the trough is spurious (see text) and it would be a waste of time to wonder about possible sociological causes. *Source: CDC-WONDER (http://wonder.cdc.gov/mortlCD9).*

Bronx county: 1999, 39; 2000, 123; 2001, 23; 2002, 50 (CDC-WONDER). As these numbers are of the same order of magnitude as the numbers of suicides or the numbers of murders, this raises some doubts about the reliability of suicide and/or murder data.

5.4 Interactions between Japanese population and occupation forces

Before presenting some evidence about the occupation of Japan, we describe the context.

5.4.1 Context

The occupation of a country by a foreign army is a phenomenon which has occurred repeatedly in the course of history. Examples abound: occupation of Belgium and the Netherlands by the 20,000 troops of the Duke of Alva in 1567; occupation of Algeria by France in 1830; partial occupation of France by the troops of Bavaria and Prussia in 1870–1; occupation of Egypt by Britain in 1882; occupation of the Philippines by the United States in 1898; occupation of the Ruhr by Allied troops in the wake of World War I; occupation of Germany and Japan by Allied troops in the wake of World War II. Some comparative data such as the numbers of troops of the occupation army and the numbers of fatalities suffered during the occupation can be found in Roehner (2004, pp. 286–7). Occupation episodes provide an ideal testing ground for social cohesion.

Such episodes share many common mechanisms and features.

- The occupation troops are a foreign body in the sense that they are not subject to the laws and tribunals of the country. Soldiers and officers who commit crimes against the population are tried by courts martial. Even when severe sentences are meted out, the offenders are often reprieved and pardoned by their commander-in-chief. As a result occupation troops develop a feeling of impunity.
- Occupation troops have to be lodged, fed, kept warm and entertained. In a country which has been heavily bombed, like Germany or Japan in 1945, the requisition of the remaining buildings for the housing of headquarters, officers and troops puts the population in a difficult situation. Moreover, the requisition of coal and oil leads to great hardships in wintertime. Under the heading of entertainment one must also include prostitution and semi-prostitution. By definition an army is composed of young males whose needs in this respect cannot be completely ignored. Overt prostitution can be organized by setting up "entertainment inns" reserved to occupation troops. Semi-prostitution hinges on food, money and other favors that women can obtain from soldiers or officers. Needless to say, such relations are a source of friction between troops and the male population.
- Another cause of friction is the question of reparations. In Germany and Japan it took the form of the dismantling of factories and their appropriation by the nations participating in the occupation.
- The trial of political leaders and officers, the former for having started the war, the latter for being responsible for war crimes, was carried out for the first time after World War II. After World War I, there was much talk, especially in the United States, about

trying former emperor Wilhelm II. Eventually, however, the plan was dropped when it became clear that the Netherlands would not extradite him. As a result of the trials that took place in occupied Germany and Japan, about 500 Germans and 920 Japanese were executed (*Quid* 1997, p. 1088; Dower 1999). It seems clear that, no matter how justified they may have been, these trials antagonized former military personnel. In the case of Germany we know that this resulted in a number of bombing attacks against denazification tribunals.

Almost all occupation episodes reveal incidents and clashes between occupation troops and population. This is hardly surprising on account of the causes of friction listed above. Yet there is one conspicuous exception, namely the occupation of Japan. Writing in 1955 in the Swiss newspaper *La Gazette de Lausanne* (November 4) the renowned journalist Tibor Mende notes that there was *not a single* serious incident between Japanese and Allied troops during the whole occupation period. Similar statements can also be read in more recent publications written by historians such as Manchester (1978), Finn (1992) or Dower (1999). In the light of what we said earlier in this chapter about the process of replication, this exception poses a serious problem. In so far as this exception concerns a major country and an important historical episode it cannot be easily dismissed. This conundrum can be solved in two ways:

- One can try to identify factors which may explain why Japan should be an exception.
- One can try to question the data on which the present opinion is based. Is the historiographical record of this period really convincing or are there omissions?

In many ways Japan is a very special country. Thus, if one favors the first explanation there are countless factors which can be invoked. One factor which is frequently mentioned is the role of Emperor Hirohito. He asked all Japanese to collaborate with the Allies and therefore, it is argued, the Japanese people had no other choice than to comply. From a historical perspective this story is not really convincing, for Hirohito was not the demigod described in some historical accounts. Furthermore, on numerous occasions his explicit will was opposed by various groups of Japanese. Let us briefly recall these two points (Behr 1989).

- It is often written that before 1945 the Japanese had never been able to see their emperor. This is not true. In November 1928, Hirohito traveled by train from Tokyo to Kyoto where his marriage was to take place. The journey took two days because of frequent halts which gave Hirohito the opportunity to greet the thousands of Japanese massed along the railroad. Moreover, Hirohito attended military parades and official dinners just like any other head of state.[5]

---

[5] For instance, on June 18, 1941, he attended a dinner in honor of Wang Ching Wei the puppet prime minister who ruled the northern part of China then occupied by Japan.

- In Japan, emperors have little say in political matters. Emperor Taisho (1912–26) was a great admirer of Germany; yet Japan declared war on Germany in World War I. Emperor Hirohito was fond of England which he visited in his twenties; he kept close personal contacts with the royal family; this did not prevent Japan from declaring war on Britain.
- During a failed insurrection in February 1923, the insurgents killed Hirohito's former personal chamberlain. On November 14, 1930 the Prime Minister was assassinated, on February 7, 1932 the Minister of Defense was murdered, on May 15, 1932 Prime Minister Inukai was killed. On several occasions, Hirohito had condemned these plots and it is therefore hard to imagine that the insurgents could convince themselves that their murders fulfilled the will of the emperor.

Of course, this short discussion does not disprove the first explanation. In fact, Japan has so many idiosyncrasies that it is an impossible task to refute this argument. Moving now to the second explanation, we face a difficult task for, as we will see, most sources which would tell us what happened are not yet accessible (Table 5.4).

5.4.2 Evidence

On September 16, 1945, that is to say a few days after the beginning of the occupation, General MacArthur imposed severe censorship. It was maintained at least until the beginning of the Korean War in June 1950 as illustrated by the following news published in the fall of 1949 in the *New York Times*: "A publisher, S. Morioka, and a professor, I. Oyama, are charged with violations of censorship policies" (September 9, p. 12; September 29, p. 15); "Six Japanese were sentenced to 5 years hard labor for spreading rumors derogatory to occupation forces" (August 31, 1949, p. 15).

For this reason the news published in Japanese or foreign newspapers cannot give anything but a fairly incomplete picture; to get a more accurate one we need other sources. Table 5.4 lists several official sources which, if available, would be able to give qualitative as well as quantitative information. The provost courts were set up in early 1946 on the model of military courts martial of which they followed the principles of law and rules of procedure. A statement made one year later on March 25, 1947 by the Headquarters of the Eighth Army in Yokohama informs us that during the year March 1946–March 1947 more than 12,000 Japanese people were tried by United States provost courts (i.e. on average 1,000 per month) for crimes committed against Allied troops and property or for violations of censorship regulations. On average 90 percent of those tried were convicted and the average prison term was 6 months. The sentences that could be inflicted by provost courts were limited to 10 years hard labor. Longer terms or

Table 5.4 *Availability of historical records about the occupation of Japan*

Type of data	United States	UK	Australia	New Zealand
Roll of Honor	NA	A	A	A
Provost court trials				
Statistics	~ 10%		NA	A
Records	~ 3%		NA	NA
Military Commission trials				
Statistics	NA	NA	NA	NA
Records	NA	NA	NA	NA
Court martial trials				
Statistics			NA	
Records			NA	
Claims for indemnities by Japanese				
Records	NA	NA	NA	NA
Investigation reports on fires				
Records	NA	NA	NA	NA

Notes: The term "Statistics" means number of cases per month whereas the term "Records" means detailed accounts of the trials or of the claims. A means "available", NA (not available) can mean two things: (i) the archivists who are in charge of such material do not know where it is located; (ii) the material is well located but the archives are not open to historians. When a percentage figure is given it means that only a fraction of the data are available. To the occupying powers mentioned in the table one should also add the Soviet Union who occupied the South of Sakhalin and the Kurile Islands (see in this respect Sevela (2001)) and Indian forces who took part in the occupation until July 1947. The "Roll of Honor" is a list of fallen soldiers and officers which comprises the dates and sometimes the causes of the death; for our purpose the names are of course irrelevant. Provost courts and Military Commission courts tried Japanese people who had committed misdemeanors, misdeeds or crimes against occupation troops. The difference was in the sentences; provost court sentences were limited to 10 years hard labor, whereas Military Commission courts could inflict heavier sentences including the death penalty. Based on data for the BCOF, the number of courts martial was of the order of 10 per month and per 10,000 troops (Brocklebank 1997). We know that about 10,000 Japanese civilians were tried by provost courts in the first year of their existence (March 1946 to March 1947), but we do not know how many Japanese were tried by Military Commissions for crimes against occupation troops. From Japanese sources, one knows that about 10,000 Japanese households obtained indemnities from their government for physical damage suffered as a consequence of the occupation (4,339 cases concerned deaths); these data were published in Tsurumi (1961) but to our best knowledge no detailed study of the circumstances of these casualties has been undertaken. There were about 250 fires a year in Allied installations; detailed investigations were made into their causes but until these reports become available we will ignore what proportion of these fires were arsons.
Sources: First year of operation of provost courts: *New York Times*, March 25, 1947 (p. 4); occupation indemnities: Tsurumi (1961, pp. 579, 583); for availability of documents: communications from archivists at NARA (national American archives), AWM (Australian War Memorial), AHU (Australian Army History Unit), National Archives of New Zealand.

death sentences could be inflicted by Military Commission courts. Unfortunately, it has not yet been possible to find statistical data about the number of trials before these courts.[6]

The number of fatalities among occupation troops provides a useful estimate of the intensity of the incidents in which they are opposed to the population. For the purpose of establishing comparisons between different cases it is best to express fatality numbers in normalized form. The number of deaths per month and per 10,000 troops is a convenient normalized variable.[7] For the sake of brevity we will refer to it as the normalized fatality rate (NFR). What is the range of variation of this rate? This is a question which one needs to address because in any army there are fatalities due to illness or accidents. According to British defense statistics (*National Statistics of the United Kingdom 2005*, p. 20) in time of peace for a force which is stationed at home or in friendly foreign territory, the normalized fatality rate is 0.70 per month per 10,000 troops. At the high end of the spectrum, one can mention the rate of $25/(\text{month} \times 10^4)$ for American troops in Vietnam in 1968; such a rate characterizes a situation which is intermediary between occupation and open warfare.

British Commonwealth occupation troops who died in Japan were buried at the Yokohama War Cemetery.[8] The annual numbers of fatalities are represented in Fig. 5.5. It can be seen that during the whole period the normalized rate is above the threshold of $0.70/(\text{month} \times 10^4)$; in 1948–9, that is to say before the outbreak of the Korean War, it is on average equal to 2.2. On the contrary, the caskets of American troops who died in Japan were shipped back either to the Philippines, to Hawaii or to cemeteries in the continental part of the United States. The National Personnel Records Center – Military Personnel Records (NPRC–MPR), located in the St. Louis suburb of Overland, Missouri, holds military records of deceased military but these archives are not accessible to historians. This is why we do not yet know the number of American servicemen who died in Japan during the occupation.

Although numbers matter, they do not tell us much about what really happened. Figure 5.6a tells us somewhat more. First, it confirms what one already observes in Fig. 5.5, namely that the rate of incidents increased between 1948 and 1952. This is also corroborated by the curves in Fig. 5.6b. Second, we can observe that the total number of assaults against BCOF personnel is about five times smaller

---

[6] Military Commission courts also tried war crimes; almost all the information which is available about the activity of these courts refers to war crimes.

[7] 10,000 is approximately the strength of a division; in fact, depending on circumstances the strength of a division can be somewhere between 8,000 and 18,000; usually the strength is reduced in peacetime.

[8] Except for an undetermined number of dead whose remains were sent back to their home country. For Australian troops this number was of the order of 60. I am grateful for this information to Mr. Ron Orwin, secretary of the Executive Council of Australia of the British Commonwealth Occupation Force.

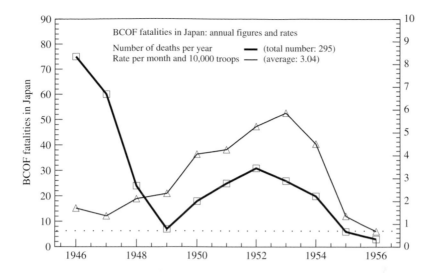

Fig. 5.5 Soldiers, officers and civilians of the British Commonwealth Occupation Force (BCOF) who died in Japan. Solid line: number of members of the BCOF who died in Japan and are buried at the BCOF cemetery in Yokohama; this count underestimates the number of fatalities because the remains of some were shipped back to their home country; for instance, about 60 Australian servicemen who died in Japan are buried in Australia (for the other BCOF countries, we do not know). Thin line: fatality rate per month and per 10,000 troops. For the purpose of comparison it can be noted that the fatality rate was about $4.5/(\text{month} \times 10^4)$ for the American force occupying Iraq in 2003–5. After the beginning of the Korean War there were continuous movements of troops between Korea and Japan; as a result the number of Commonwealth troops stationed in Japan was not well defined. *Source: BCOF Council of Australia. Many thanks to Mr. Ron Orwin for his help.*

than the number of fatalities (42 against 209 for the same time period 1946–52). So even if all assaults had been murders (which was certainly not the case) one still has to explain the high number of fatalities. Although, as already mentioned, one cannot count on newspaper articles to give a realistic picture, nonetheless they can provide some hints. Here are some excerpts, mostly from the *New York Times*.

- There were frequent **fires** in Allied headquarters, barracks or warehouses. In the three months January–March 1947 there were 67 fires in the American zone, which puts the annual rate at over 250 fires. Below are a few excerpts. (i) "Three officers are killed in a fire at camp Schimmelpfennig" (January 9, 1947); the French newspaper *Le Monde* gave the following brief account: "The Japanese once again attempted to set American barracks on fire." (ii) "A barrack building was burnt down by Japanese arsonists at Yamaguchi in the New Zealand zone of occupation" (April 4, 1947, Brocklebank 1997, p. 77). (iii) "A Japanese, Kumano Mitsuji, is arrested by New Zealand troops and tried

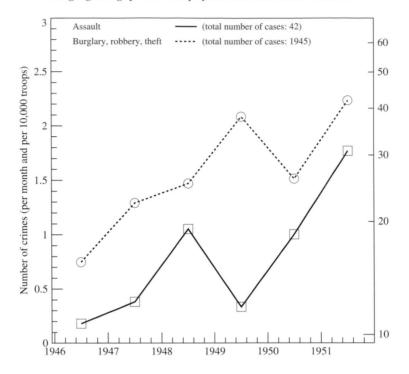

Fig. 5.6a Japanese civilians against BCOF personnel. The left-hand scale refers to assaults, while the right-hand scale refers to burglaries and robberies mainly due to the great scarcity that existed in Japan in the first post-war years. The main interest of this curve is to provide a consistency check on the curve for the number of assaults by Japanese on BCOF personnel. *Source: Carter 2002, p. 309, thesis submitted to the Australian Defence Academy at the University of New South Wales.*

on charge of arson of occupation forces barracks." (National Archives of New Zealand, WA-J76/1) (iv) "During the night of April 24–25, 1948 three fires occurred in the Headquarters of the British troops in Kure; the first started at 11:40 pm, the second at 1 am, the third at 3 am" (Australian War Memorial Archives, AWM52 18/1/11).

The acknowledgment that the fire was due to arson was extremely rare. Most often fires were attributed to defective wiring even though most fires began during the night, that is to say at a time when all lights and electric equipment were turned off.

- **Sabotage** actions: (i) "Seven civilians were seized for carrying grenades and other arms" (*New York Times*: November 22, 1945, p. 16). (ii) "On 6 April 1946, working under cover of darkness and avoiding detection by hourly military police, saboteurs destroyed thousands of dollars worth of communications and aircraft equipment at Chofu Airfield 24 km west of Tokyo in what is believed the first organized Japanese

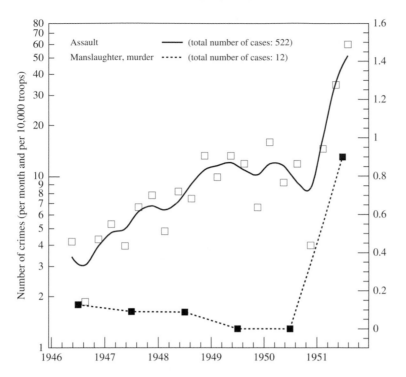

Fig. 5.6b Crimes committed by BCOF troops against Japanese people. These numbers are based on BCOF statistics. The ratio of the number of assaults by BCOF personnel on Japanese to the number of assaults by Japanese on BCOF forces is $522/42 = 12.4$. *Source: Carter 2002, pp. 287, 289, thesis submitted to the Australian Defence Academy at the University of New South Wales.*

violence against occupation forces property" (*New York Times*: April 6, 1946, p. 4). (iii) "On 21 January 1947, a Japanese was arrested in Kyoto on arson charges. He allegedly tried to set fire to a US Army counter-intelligence office at Saga. The Army said the incident was one of the first cases of such action against Allied forces since the beginning of the occupation" (*New York Times*: January 21, 1947, p. 12). (iv) "On 15 July 1947, a group of 6 Japanese were arrested and tried by a provost court for the possession of dynamite, detonator, and fuse. Other arrests under the same charge took place on 21 June and 6 September 1947" (New Zealand Archives WA-J 76/1). (iv) "On 10 October 1948 a Corsair aircraft of a New Zealand squadron was set ablaze at Bofu airport" (Brocklebank 1997, p. 208).

- **Clashes** between Japanese and Allied troops. (i) "Private R. C. Young who caught Japanese looting a warehouse was bayoneted to death. One of the three Japanese was sentenced to death and hanged on 18 May 1946" (*New York Times*: January 26, 1946 p. 6; May 18, 1946, p. 4). (ii) "During the month of February 1948, the GHQ of the First US Corps lists 15 acts of violence against American troops, four of which involved firearms" (Eichelberger Papers, volume 33, p. 2008).[9] (iii) "While walking near the Kotobuki bridge, two Australian soldiers were assaulted by some 30 Japanese; they had to be sent for medical treatment" (August 16, 1948 AWM52 18/1/11). (iv) "A mob attacked occupation cars; a GI was hurt" (*New York Times*: July 10, 1949, p. 22).

- **Damage suffered by Japanese people**. Although there were some criminal cases,[10] most of the damage suffered by Japanese was probably due to reckless behavior. This statement can be illustrated by two kinds of incidents. (i) In 1945 and 1946 the Japanese people faced very difficult conditions including semi-famine; a number of them were killed when caught trying to break into Allied warehouses. The following story is typical of this sort of incident. "A GI shot and killed a Japanese at Matsuyama (south of Hiroshima). The dead was one of about 1,000 civilians who stormed a bivouac area which was being vacated to pick over the left-behind trash. One guard fired into the ground several times and when that had no effect, he fired into the mob" (*Pacific Stars and Stripes*: December 24, 1945, p. 1). (ii) Another illustration was the frequency of traffic accidents. In an official Australian account we read that in 1948 about 45 Japanese were killed as a result of traffic accidents involving BCOF vehicles (AWM52 18/1/11). In proportion to the small number of vehicles of the BCOF force this is a very high figure. Indeed, in 1948 the BCOF was reduced to the Australian contingent and this force comprised less than 10,000 troops. For a country of 100 million adults like Japan such a rate would imply 450,000 pedestrians killed in vehicle accidents every year. From a document published in a book by Shunsuki Tsurumi (1961) one learns that 4,339 Japanese households obtained indemnities from the Japanese government for deaths suffered as a result of the occupation. Unfortunately, we do not know the causes of these deaths, but it is reasonable to think that some of them were due to traffic accidents. If the rate of 45 per year for the Australian force is extended to the total occupation force and to the whole duration of the occupation, one gets a figure of $45 \times 10 \times 5 = 2,250$. If this estimate is correct, it means that about one-half of the 4,339 deaths should be attributed to traffic accidents.

Together with the previous graphs this enumeration shows that the view of the occupation presented by Tibor Mende and which still prevails nowadays is fairly inadequate. It also suggests that the incidents that we mentioned are only the tip of the iceberg.[11] Once the archives are opened to historians we will be able to see this episode in a perspective in which it will no longer appear as a disconcerting exception.

---

[9] The Eighth Army which occupied Japan was composed of the First, Ninth and Tenth Corps; if one admits a strength of 35,000 for the First Corps the 15 assaults per month correspond to a rate of $4.3/(\text{month} \times 10^4)$.

[10] e.g. "Private Stratman Armistead, 32, was sentenced to be hanged for having murdered 4 Japanese with a hammer. The sentence is subject to review by Lieutenant General Eichelberger, commander of the Eighth Army" (*New York Times*: April 16, 1948, p. 7).

[11] A fairly detailed chronology of all the incidents mentioned in the sources which are available so far can be found on the author's website: http://www.lpthe.jussieu.fr/~roehner.

Table 5.5 *Movements of the Earth which may be observable with a high accuracy Foucault "pendulum"*

Rotation	Period (day)	Angular velocity (degree/24 h)	Precision required (%)
1 Rotation of the Earth on its axis	1	360	
2 Rotation of the Earth around the Sun	365	0.986	0.20
3 Precession of the equinoxes		1.4×10^{-3}	4×10^{-4}
4 Rotation of the Solar System around the center of the Galaxy	73×10^9	1.3×10^{-11}	4×10^{-12}

Notes: The angular velocity column gives the angular deviation in 24 hours of the plane of a Foucault pendulum located at the North Pole as observed by a terrestrial observer. The last column gives the precision with which one must measure the deviation in order to be able to detect the rotation mentioned in the same line of the table. The accuracy of standard Foucault pendulum experiments is between 0.5% and 1% and is therefore too low for effects 2, 3 and 4 to be observable. The so-called precession of the equinoxes is a slow change in the direction of the axis of rotation of the Earth, an effect which is similar to the phenomenon by which the axis of a spinning top "wobbles" when a torque is applied to it. In addition to the movements listed in the table, our galaxy is also moving toward the Andromeda galaxy with a relative velocity equal to 220 km/s (which is about seven times the speed of the Earth on its orbit around the Sun). However, at time of writing, it is not clear if this movement is also a rotation around some (still unknown) center.

5.5 Conclusions and perspectives

Adopting the replication procedure as a standard requirement in the social sciences would truly represent a watershed for it would lead to a number of robust regularities which would give the field firm foundations. But replication goes even further than that. In the course of time as the accuracy of measurements is progressively increased there comes a moment when our understanding becomes inadequate. A classic example is the fact that the orbit of the planet Mercury could be explained by Newtonian mechanics until more accurate measurements made clear that another effect was at work. This kind of remark may seem easy to make with the benefit of hindsight. Here is a more speculative illustration. The best accuracy achieved in the experiment of the Foucault pendulum remains fairly low; is it possible to imagine a Foucault pendulum whose precision would be 10^{-5} % or 10^{-6} %? There seems to be little hope of achieving such an accuracy with a pendulum because there are too many sources of noise, but there may be other devices based on the same principle, e.g. a vibrating rod whose extremity is clamped in a chuck or perhaps vibrating (non-colliding) molecules. Assuming that a higher accuracy can indeed be achieved, what could we learn from such an experiment? The Foucault pendulum is said to oscillate in a plane which is fixed relative to "the distant stars". Obviously, this is a fairly vague statement. Do these distant stars refer to stars which are in the vicinity of the Earth, or to stars in distant galaxies? A high accuracy Foucault "pendulum" would provide experimental answers to such questions. Table 5.5 lists movements of the Earth which may become observable with such a device.

In this chapter we did not discuss cases in which the data are purposely distorted or twisted. This kind of distortion is uncommon in physics but fairly common in human societies. This is why we devote a separate chapter to this problem. Needless to say, in time of war, the extent to which news is corrupted by deception becomes much greater. This effect is well documented in Knightley (2004). In the preface of a book about the occupation of Germany one finds the following sentence: "German witnesses who demonstrated a historical perspective by seeing the good side of the victors and the bad side of the losers have been accepted as the more credible" (Peterson 1990). Statements as explicit as this one are rare, even in books published 40 years after the end of the war. Yet the example of the occupation of Japan shows that it is easier to follow mainstream views than to question them.

Part II
Macro-interactions

The chapters in Part II consider different kinds of *macro-interactions*, by which we mean interactions which take place between states or which act at the global level of societies. What makes the study of macro-interactions comparatively easier than the study of interpersonal micro-interactions is the fact that they are strong, persistent and their effects are often clearly visible. On the contrary, micro-interactions are much weaker, mostly of shorter duration and their effects are often masked by those of macro-interactions.

The motivation and rationale for first studying macro-interactions can best be explained by way of a parallel with physics. It is well known that by sorting molecules according to their velocities, a device known as a Maxwell demon would be able to make a system work in violation of the second law of thermodynamics. Naturally, it would be easy to imagine other kinds of Maxwell demons whose intervention would result in the violation of other physical laws. With such demons at work in our laboratories it would be impossible to study physical systems in a meaningful way, at least until their intervention could be identified and discounted. Fortunately, Maxwell demons do not exist in physics, but in the study of social phenomena we must indeed face this kind of difficulty. In our societies there are numerous Maxwell demons who distort and influence social phenomena; if for some reason they cannot tamper with the phenomena themselves, they will try to alter and misrepresent their historical accounts. Until we are able to identify and discount their influence it will be impossible to collect evidence in a meaningful way.

Surprisingly, there is little awareness among social scientists about these obstacles. More precisely, there is a misplaced perception. Often, social scientists think and suggest that their theories, statements and predictions may alter the behavior of social agents. In our opinion this effect has been largely overstated. In fact, the influence of theories on the behavior of social agents has rarely (if ever) been established.[1] On the contrary, we know for a fact that there are numerous organizations whose missions and objectives consist in acting as Maxwell demons. Among these groups one can mention intelligence agencies, public relations companies, think tanks, foundations or non-governmental organizations, religious sects, and so on. As some of these groups run on budgets of several billion dollars a year, it would be unrealistic to think that their action is negligible.

Of course, disinformation is nothing new, but the problem is of greater importance nowadays because the power of the media has expanded considerably.

---

[1] A striking counter-example was provided by the NASDAQ bubble in the 1990s. For over a century, one of the best established empirical rules about stocks has been the fact that the ratio P/E = (price of a share)/(earnings per share) remained between 10 and 25. In the late 1990s this ratio climbed to 200 but the buying spree continued unabated. Apparently, the "infectious greed" of investors, as the Chairman of the Federal Reserve Alan Greenspan called it in 2002, was stronger than the laws of economics and the warnings of lucid economists.

The Jeffersonian ideal of an informed citizen as an indispensable ingredient of democracy is more and more challenged by an increasingly sophisticated opinion-molding apparatus. The forthcoming chapters provide instances of situations in which it was possible to identify and to some extent discount the influence of Maxwell demons. Although this can be done only in a few cases the broader objective is to draw the attention of social scientists to this issue.

6

Shaping the Zeitgeist

The Zeitgeist, a word of German origin which literally means the spirit of the time, refers to a set of attitudes, beliefs and ideas that, at one point in time, are accepted uncritically because they appear to be shared by almost everybody. The best way to convince oneself that well-accepted ideas are in fact a product of their time is to look back at past centuries. Two examples from eighteenth-century England may illustrate this point. (i) At that time it was held as self-evident that in order to become an English "citizen" a foreigner had to embrace the Anglican faith. (ii) The fact that trading slaves, provided they were not Christian, was considered a legitimate business for decent Christian merchants is attested by many documents from religious pamphlets to court rulings. Once we have accepted the notion that most of our beliefs are shaped by the society in which we live, we would like to understand how these views, beliefs and convictions are formed. In fact, this question has far reaching consequences in various fields, from marketing to political campaigns to speculation frenzies. In recent times the issue of consensus formation has attracted the attention of econophysicists and several models have been proposed, e.g. Behera and Schweitzer (2003), Quentin and Bouchaud (2005), Stauffer (2003). One of the main difficulties in the study of consensus formation is that we do not know how to distinguish between the action of small and big players. What we mean by these expressions can best be explained through an analogy with stock markets. A small transaction, e.g. the sale of 1,000 shares by a petty shareholder, is a transaction that does not substantially affect the price of the stock whereas a big transaction, e.g. the sale of 2 million shares by an investment fund, will markedly shift the price level.[1] One may say that micro-players can only *react* to price fluctuations whereas macro-players can *shape* them. The same distinction can be made regarding consensus formation. Some actors have only a passive role whereas macro-players are able to shape the consensus.

---

[1] This is why such transactions are often settled in what is called the upstairs market, as opposed to the open market, at prices which are different from current market prices.

Naturally, this raises the question of how macro-players can be defined, how they work and to what extent they succeed in shaping the spirit of the time. It is to these questions that the present chapter is devoted. We investigate three specific cases of consensus formation. The first one concerns the microeconomic level of marketing campaigns. The second case investigates consensus formation in broad social issues; it leads us to study the role of public relations companies. The third case concerns global socio-political issues such as the emergence of the neoliberal creed. These cases differ in their themes but more importantly they differ in their time scales: the first episode lasts a few months, the second covers about one decade and the third spans several decades.

6.1 Marketing campaigns: shaping the response of consumers

It has been argued that statistics on movie attendance may provide a way to probe the level of social interconnection among moviegoers.[2] The reasoning goes as follows. Assume that in a population of N people a proportion H hear about a new movie and decide to watch it. Thus, in the first week after release, the movie will be watched by $A_1 = NH$ people. Assume further that a proportion p of them loved it to the point of recommending it to a number of friends. If each of them has F friends and if a proportion q follow this advice, the number of moviegoers in the second week should be $A_2 = A_1(pFq)$.[3] Thus, if all other quantities are known, it should be possible to infer F from A_1 and A_2. If the same argument is repeated for subsequent weeks, one gets for the nth week $A_n = A_1(pFq)^{n-1}$. If $pFq < 1$ the attendance curve will be decreasing geometrically. This is indeed what is observed for most movies (Fig. 6.1a); roughly, every week the attendance is divided by a factor of $pFq = 1.5$, which implies that $F \sim 1.5/pq$. However, in a few cases one observes an attendance which *increases* in the course of time (Fig. 6.1b). How can one explain this effect? Our main interrogation is whether this attendance curve should be explained by one of the factors (such as p, F or q) which describe the responses of moviegoers or rather by the effects of the marketing campaigns. It is to this question that the present section is devoted.

Before we discuss this point let us have a closer look at first week attendance. Observation shows that there is a close connection between the marketing budget B of a movie and the audience A_1 in the first week (Fig. 6.1c):

$$\text{Opening week receipt} = 0.25\,(\text{Marketing budget})^{\alpha} \qquad \alpha = 1.3 \pm 0.4$$

---

[2] The same argument can be made for book sales; see in this respect the pioneering paper by Sornette *et al.* (2004); a more detailed version can be found in Deschâtres and Sornette (2004).

[3] For the sake of simplicity we ignore people who initially wanted to watch the movie, were unable to watch it in the first week and postponed attendance to a subsequent week.

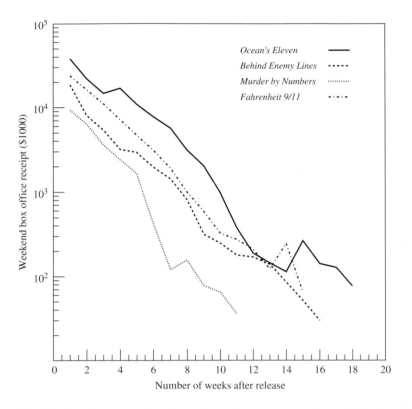

Fig. 6.1a Weekend receipts for four films. The downward trend displayed in this graph for the four selected films reflects a pattern which applies to 90% of films. Roughly, in each week after release the receipts are divided by 1.5. *Sources: http://www.boxofficemojo.com/movies; http://imdb.com (Internet Movie Database).*

Most often the marketing campaign takes place in the two or three weeks before release. Understandably, the magnitude of the marketing campaign determines the fraction H of the population N which hears about the movie and decides to watch it.

We now turn to the few movies whose attendance curve is increasing in the course of time. This situation is illustrated by the movie *Napoleon Dynamite*. Aimed at an audience of teenagers, the movie was released on June 11, 2004. To begin with, one should observe that it differs from other movies in several features.

- Usually, the marketing budget of a movie is smaller than its production budget. However, for *Napoleon Dynamite* the marketing budget represented 25 times the production budget.
- For an average movie the receipt in the opening week represents about 30% of the total receipt; for *Napoleon Dynamite* it represented less than 4%.

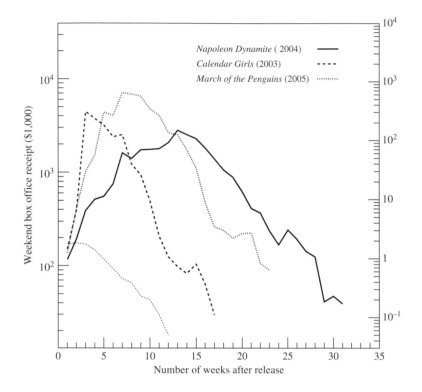

Fig. 6.1b Humpback shaped receipt curves. For a small number of films the receipts increase after the opening week. For *Calendar Girls* the graph shows two separate curves: one for attendance in the United States (thick broken line, left-hand scale), the second for attendance in the UK (thin broken line, right-hand scale labeled in millions of pounds), the curves for the other two films concern receipts in the United States. One might be tempted to interpret these curves as reflecting a spread-the-word process; however, evidence about the organization of the marketing campaign suggests that these humpback receipt curves were planned by the distributors. *Sources: See Fig. 6.1a.*

These two figures immediately suggest that the marketing campaign of this movie was very unusual. A closer look reveals the following features.

- In contrast to standard movies which open in as many as two or three thousand theaters, *Napoleon Dynamite* was released in only a few dozen. In subsequent weeks the number of theaters was progressively increased to reach 179 screens six weeks after release. What happened in the remaining weeks will be discussed in a moment.
- During these first six weeks, 100,000 printed T-shirts were distributed; frequent viewer cards were given to encourage children to watch the movie a second, third or fourth time. Moreover, viewers who came back with a friend received little gifts such as pins or lip balms.

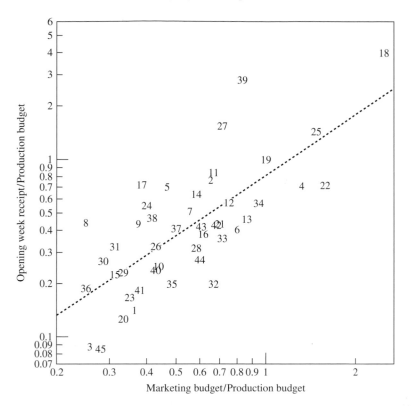

Fig. 6.1c Relationship between opening receipts and marketing budget. The correlation of the scatter plot is 0.73. The slope of the regression line of the log-log plot is 1.1 ± 0.3 which means that the two ratios are almost proportional. Three of the outliers which are above the regression line are *Spiderman* (8), *Shrek 2* (27) and *The Passion of The Christ* (39); two of the outliers located below the regression line are *New York Minute* (32) and *The Alamo* (45). In this sample the highest marketing budget / production budget is *Farenheit 9/11* (18). *Sources: See Fig. 6.1a.*

- The film producers released comments to the media saying that they were thrilled with the enthusiastic response the film had received and emphasizing that fans had seen the film three times or more.
- Games and animations were organized on the website of the film which allowed the producer to claim that "over 25,000 people are competing to become President of the *Napoleon Dynamite* Fan Club."

The general objective of the campaign was to create the impression that the movie had generated genuine enthusiasm among spectators and that the number of viewers had been increasing as a result of a collective mania effect.

On July 24, 2004, after six weeks of this grass-roots campaign, a kind of second opening was staged, which included the addition of a five-minute epilogue in

order to give the impression that the movie was still in the making as a result
of an interaction between the production team and the reactions of the viewers.
In this second phase the number of screens was increased from 179 in week 6
to 389 in week 7; the expansion continued in subsequent weeks until there were
1024 screens in weeks 15 and 16. This strategy seemed to work well in the sense
that there was a marked peak in attendance 13 weeks after the film was released
(Fig. 6.1b).

We now come back to our initial question about exogenous versus endogenous
forces. Figure 6.1c is crucial in this respect. The connection between marketing
budget and attendance in the first week shows that attendance levels can be
controlled exogenously by advertisement campaigns without any spread-the-word
effect (such effects can possibly play a role only *after* the first week). Thus, if
instead of concentrating the marketing campaign in the weeks before release (as
is done usually), the distributors set up a campaign which covers the first two
months after release, this campaign will be able, week after week, to draw an
attendance in proportion to the marketing outlays. In other words, one suspects
that the exogenous effects of the marketing campaigns have had a predominant
role. To find out more precisely, one would need to know detailed figures of
marketing outlays in the weeks after opening.

Figure 6.1b shows two other attendance curves which display a marked peak.
Were these peaks also engineered by a clever marketing campaign? For *March
of the Penguins* the comments made on the Internet insisted on the fact that the
fancy was unexpected and genuine: "Who knew that the penguins in Antarctica
could be so cute, charming and completely captivating?"; "It is the first time that
a documentary reaches such a broad audience"; "We have an 80 million receipt
to date [i.e. by the end of November 2005]: not bad for a French made bird film."
However, before one can conclude that this effect was really endogenous one
needs to know the scale of the marketing campaign. Warner Independent Pic-
tures, the distributor, spent nearly $30 million to promote the film in the United
States. Such a marketing budget is in line with standard Hollywood productions
(for instance the marketing budget of *Catwoman* was $35 million) and is quite
exceptional for a documentary. For the opening weekend (June 26, 2005[4]) the
film was released in only 4 theaters, in 20 in the second week, 64 in the third,
until eventually being screened in 2,500 theaters in the twelfth week. Was this
planned in advance or were additional copies made only as they were needed?
It is difficult to know, for the the information available about the marketing
campaign of *March of the Penguins* is less detailed than for *Napoleon Dynamite*. It can be noted that in France the film receipts did not exhibit any peak:

---

[4] It was almost, one year later, the same release date as *Napoleon Dynamite*.

the data for the five weeks following the release on February 1, 2005 were as follows (in millions of euros): (i) 2.5 (408 screens), (ii) 2.0 (457 screens), (iii) 1.7 (472 screens), (iv) 1.4 (473 screens), (v) 0.96 (461 screens). Thus, despite a number of screens which was slightly increasing, weekend receipts decreased steadily.

Movies seem to be one of the few products for which reliable data about marketing budgets are made public.

In the next section, in which we try to understand the effect of public relations campaigns, we will not be able to rely on outlay data.

6.2 Public relations campaigns: example of cell phones in cars

In the previous section the objective of the campaign was to promote a specific product. We discussed the case of movies but similar techniques are used in the promotion of other products. In contrast, the issue in the title of the present section does not refer to a specific brand, but has great implications for the whole economic sector of wireless carriers and cell phone companies. According to an estimate given by the *Wall Street Journal* (July 19, 2004), 40% of the traffic on cellular phones in the United States is due to drivers, a share representing an amount of $37 billion. For the purpose of promoting their broad common objectives, wireless telecommunication companies have joined forces by forming influential associations such as the Cellular Telecommunication and Internet Association (CTIA).[5] The purpose of such associations is to represent their members with policy makers and to set up broad public relations campaigns.

The issue of whether one should use cell phones while driving is interesting because it is a case in which the evidence is particularly clear. Thanks to numerous observations carried out by safety agencies we know that using a cell phone while driving increases the risk of accident by a factor of four or five. Yet government regulations have consistently failed to take this evidence into account. In what follows we examine these two points in more detail. First, we present and discuss the evidence on risks, then we survey the responses in terms of legislation in several industrialized countries.

Cell phones have been in use since at least 1995 and radio telephones have been used for much longer, particularly by truck drivers. The question of safety implications has been actively investigated by safety agencies and by university laboratories. Several landmark investigations are summarized in Table 6.1. One of the most unquestionable observations was made in Japan. The ban on using

---

[5] Other similar associations are the National Cellular Association and the Broadband Wireless Association.

Table 6.1 *Studies on the risk of using cell phones while driving*

	Year	Country	Main conclusions
1	1995	France	Reaction time of drivers is 60% slower
2	1997	Canada	Risk of accident multiplied by 4 to 5; hands-free phones offer no benefit
3	1999	Japan	• In the 6 months before the ban there were 1,473 cell phone related accidents; in the 6 months after the ban there were 580 • In the 12 months before the ban there were 2,830 cell phone related accidents; in the 12 months after the ban there were 1,391
4	2000	UK	Fourfold accident risk during (and up to 5 minutes after) cell phone calls
5	2001	United States (Utah)	Phone conversations create distraction levels much higher than other activities such as radio, talking with passengers, etc.
6	2004	United States (Harvard)	Cell phone related accidents kill 2,000 people each year in the United States (330,000 injured)
7	2004	Sweden	No significant difference between hand-held and hands-free phones

Notes: The third column indicates the countries where the studies were carried out. It should be noted that Japan was one of the few countries where police reports gave indications about cell phone use in accidents; in most other countries this information is not recorded.

Sources: 1: *Le Monde* (December 29, 1995); 2: *New England Journal of Medicine* (February 1997) cited in *Le Figaro* (February 13, 1997); 3: *Edmonton Sun* (Alberta, Canada, February 27, 2004); 4: *The Independent* (May 6, 2000), *Daily Telegraph* (October 4, 2003); 5: *Daily Mail* (January 30, 2003); 6: *Edmonton Sun* (Alberta, Canada, February 27, 2004); http://www.ncsl.org; 7: *International Herald Tribune* (April 12, 2004).

hand-held cell phones in cars was imposed in November 1999. In the six months before enforcement there were 1,473 traffic accidents connected with drivers using mobile phones, whereas in the six months after the ban there were only 580, which represents a decrease of 60% (ROSPA 2002, p. 18).[6] The other conclusions which emerge from Table 6.1 are the following.

- The risk of accident is multiplied by a factor of four or five when a cell phone is used in a car.
- Hands-free phones offer no benefit. A study by researchers at the University of Toronto and published in the *New England Journal of Medicine* (February 1997) investigated a comparison between hand-held phones and hands-free phones. The risk was found to be four times higher for the first device and six times higher for the second. This

---

[6] Japan is one of the few countries whose accident records include the question of cell phone use.

result might seem to be fairly counterintuitive but becomes more understandable when one realizes that people with hands-free phones tend to make longer calls than people with hand-held phones; moreover, all studies show that it is not the fact of holding the wheel with one hand which is dangerous but rather the operation of dialing the number and the fact that the driver's attention is captured by the conversation.

- A study published in early 2004 by the Harvard Center of Risk Analysis estimated that drivers talking on cell phones are responsible for about 6% of all auto accidents which occur in the United States. This represents annually about 2,000 people killed and 330,000 injured (Sundeen 2004, p. 3).

At this point a few comments are in order regarding possible objections that come to mind. The main question is whether the *fourfold* ratio reported in the Canadian study is consistent with the *twofold* decrease in accident number observed in Japan.

(1) First one should note than only hand-held phones were prohibited in Japan; if all kinds of phones had been prohibited there would certainly have been a sharper decrease.

(2) The previous answer raises another question. If, as shown by the Toronto study, hands-free phones are not safer than hand-held phones, how can one explain the decrease observed in Japan? Back in 1999, only a few drivers already had hands-free sets, which means that the majority of the drivers had to stop using their phones altogether until able to purchase a hands-free set. If this interpretation is correct, one would expect the number of accidents to increase in the course of time; this is indeed confirmed by the fact that in the second six-month interval after the ban the number of accidents jumped from 580 to 811.

(3) One may wonder why conversations on the phone are more distracting than conversations with passengers. There are three explanations. (i) Passengers spontaneously stop talking when the driver faces a difficult situation, for instance during a tricky overtaking. (ii) There is evidence that phone conversations have a high emotional content; usually, people use them to say something important. (iii) Emotional conversations with passengers may also be a source of accidents but there are no data on this question because, in contrast to cell phone calls for which connection times are known exactly, one does not have any information about the conversation which took place in the minutes preceding an accident.

The evidence presented in Table 6.1 suggests that the only sound attitude would be to forbid the use of cell phones by drivers, whether hand-held or hands-free. This would save at least 5,000 lives annually in North America and Europe and a much larger number in countries with large populations such as China, India or Indonesia. This leads us to the central question: how was the information about risks circulated in different countries?

The dissemination of information by the media, road safety agencies, automobile clubs and so on can be studied in two ways: (i) by trying to survey, assess and quantify the content of newspaper articles or agency reports; (ii) by examining the

Table 6.2 *Ban on cell phone use while driving: cross-national comparison*

	Country	Year	Month
1	Austria	2002	
2	Canada	No ban	
3	Denmark	1998	July
4	Finland	2003	January
5	France	2003	April
6	Germany	2001	February
7	Italy	2003	July
8	Japan	1999	November
9	South Korea	2001	July
10	Spain	2002	
11	Sweden	No ban	
12	UK	2003	December
13	United States	No ban	

Notes: The bans concern hand-held phones only; so far no country has banned hands-free phones in spite of reliable evidence showing that using them is no less dangerous. "No ban" means that (as of July 2004) there has been no nationwide ban. In New York State, overriding Mayor Michael Blomberg's veto, the City Council has passed a ban which became effective in November 2001. However, ticketed drivers can escape the $100 penalty if they can prove that they have bought a hands-free set since they were stopped. In July 2004, New Jersey became the second state with a ban on hand-held phones; however, police may charge violators only after stopping them for another infraction. In European states, the penalty ranges from 30 euros in Germany to 60 euros in Denmark.
Sources: 1, 4, 5, 6, 7, 10, 11, 12: Agence France Presse (December 1, 2003); 2, 3, 8, 9: http://www.cellular_news.com/car_bans, http://www.nj.com/printer; 13: *Guardian* (March 23, 2002), *Miami Herald* (March 8, 2004).

legislation which was enacted to cope with the problem. The first methodology involves many hurdles because there is a great variety of publications and articles and, even more importantly, their content is difficult to assess in an objective way. The second approach focuses on the outcome in terms of new legislation. This information is presented in Table 6.2. As can be seen, by mid 2005 not a single country had banned hands-free phones. This suggests that, although potentially available, the information about their risks had not been circulated. In contrast, the information about the risk of hand-held phones had been disseminated in many countries but legislation was enacted with great differences in timing. It could be argued that legislation is more difficult to pass in centralized countries such as Britain, France and Japan than in states which have a federal structure and where legislation can be passed at state level. However, we rather observe the opposite. In the United States the process took even longer at the level of individual states than in countries where the decision had to be taken at national level.

In the case of some other life-threatening factors, the dissemination of information is much more rapid. The alert regarding Severe Acute Respiratory Syndrome (SARS), an atypical pneumonia, spread worldwide in a matter of weeks in the spring of 2003; yet according to the World Health Organization it caused only 774 deaths. Most conferences and conventions scheduled for Toronto were canceled; on April 22 the Canadian Broadcasting Corporation reported that the hotel occupancy rate in Toronto was only half the normal rate. Yet there were only 44 deaths in Canada. Clearly, the spread of the information and the consensus formation progressed at very different rates in the two cases. For SARS the time scale was a matter of weeks while for cell phones it was a matter of decades, which means a ratio of the order of 500. The natural conclusion is that normal information channels did not perform their role in the case of the cell phone issue. If one wishes to get a better understanding the best strategy is to look at similar cases. In many instances in which (i) there are life-threatening risks either for consumers or for workers, (ii) the remedy would imply huge financial losses for manufacturers, one can observe a suppression of information or a restraint in the way it is spread. One can mention the following episodes which took place over the past 40 years (the date within parentheses approximately gives the year in which the problem emerged): asbestos (1930), smoking (1950), Alar on apples (1973), transfat (1975),[7] passive smoking (1975), latex gloves (1990), genetically modified organisms (1998). More details on these questions can be found in Rampton and Stauber (2001). The appendix at the end of this chapter gives some details about the methods used by public relations companies to shape accepted views.

In the next section we consider a question which has even broader implications.

6.3 Shaping the Zeitgeist: the promotion of neoliberalism

Ideological confrontations are not uncommon in the course of history. The long-lasting antagonism between Reformation and Counter-Reformation was one example; the competition between the communist bloc and the West was another; the war of ideas between Keynesianism and neoliberalism is a third one. It would be a very narrow view to see the struggle between Reformation and Counter-Reformation as purely religious; it had vast economic and political implications. Was the confiscation of Church property not one of the first steps taken by Reformation movements? Similarly, it would be a restrictive perspective to consider the question of neoliberalism versus Keynesianism as a purely economic debate.

---

[7] See Roehner (2005c).

Rather than an intellectual controversy which could be settled by rational arguments, it is a confrontation between two creeds. This is why it provides a good
testing-ground for consensus formation effects.[8]

At the time of writing (May 2006) neoliberalism is the sole economic paradigm
in Europe and in the United States. Keynesianism or neo-Keynesianism has
become anathema. First, we describe this situation in more detail. Then we emphasize that the neoliberal ideas did not emerge with Margaret Thatcher or Ronald
Reagan. In fact, they had been actively promoted since the end of World War
II and even in the 1930s. We try to understand the forces and institutions which
took part in this process. Finally, we consider neoliberal policies in the light of
the network theory perspective.

Broadly speaking, neoliberal ideas are defined through expressions such as
"free-market ideology" or "laissez-faire capitalism". The political objectives of
neoliberals can be summarized under three headings: (i) opposition to state interference and to unions; (ii) deregulation and privatization of utility companies
(water, electricity), of transportation companies (railroads and airways), of health-
care, of higher education; (iii) neoliberalism favors the opening of foreign markets
by political, diplomatic or economic pressure; this objective is related to the first
two objectives in the sense that reducing the role of national governments to a
minimum facilitates economic penetration. We already mentioned that neoliberalism seems to reign supreme, but is it possible to give this statement a more
precise meaning? Between 1985 and 2005 there was rapid growth in the number
of neoliberal think tanks in Europe, from 4 to 180. The last figure relies on a fairly
objective definition in the sense that it corresponds to the number of think tanks
which have formed the so-called Stockholm network,[9] which has as its explicit
objective the propagation of free-market ideas. Table 6.3 gives the numbers of
affiliated think tanks for European countries. It reveals a real fervor for neoliberal
ideas even in Scandinavian countries, once the homeland of national solidarity
principles.

As we already mentioned, the active promotion of neoliberalism goes back (at
least) to the 1930s. However, a major move occurred in 1944 with the publication
and wide diffusion of a book entitled *The Road to Serfdom*. It is to this step
that the next section is devoted. It provides a good example of an attempt by
macro-players to shape the Zeitgeist.

---

[8] Some parallels with the Counter-Reformation are discussed in the Appendix.
[9] Current director is Helen Disney, a former journalist at *The Times* of London.

Table 6.3 *Regional differences in the penetration of neoliberal think tanks*

Country	N	Pop (10^6)	Think tanks per 10^7 pop	Country	N	Pop (10^6)	Think tanks per 10^7 pop	Country	N	Pop (10^6)	Think tanks per 10^7 pop
Ukraine	1	52	0.2	Georgia	1	5.4	1.9	Switzerland	3	7	4.3
Russia	3	147	0.2	Lithuania	1	3.7	2.7	Sweden	4	8.9	4.5
Belarus	1	10	1.0	UK	16	59	2.7	Kosovo	1	1.9	5.3
Germany	8	82	1.0	Norway	1	4.3	2.3	Ireland	2	3.6	5.6
Spain	4	39	1.0	Albania	1	3.5	2.9	Czech Rep.	6	10	5.8
France	7	58	1.2	Greece	3	10	2.9	Belgium	6	10	5.9
Poland	5	39	1.3	Hungary	3	10	2.9	Bulgaria	5	8.5	5.9
Italy	10	58	1.7	Austria	3	8.1	3.7	Croatia	3	4.5	6.7
Romania	4	23	1.8	Denmark	2	5.2	3.8	Estonia	1	1.5	6.7
Netherlands	3	16	1.9	Finland	2	5.1	3.9	Macedonia	2	2.1	9.5

Notes: The column after the name of the country gives the number N of think tanks belonging to the Stockholm network which is a confederation of think tanks of neoliberal inclination. The columns under the heading "Pop" give the populations of the countries in millions; the columns labeled "Think tanks" give the number of think tanks per 10 million population. Several of the countries which before 1990 belonged to the Eastern bloc are characterized by high think tank rates. The purposes of these think tanks can be illustrated by an excerpt from one of them, Avenir Suisse: "Designed after the Anglo-Saxon model, Avenir Suisse was founded in 1999 by fourteen internationally operating Swiss companies. By communicating scientific results in clear-cut terms to a large public, Avenir Suisse enhances the circulation of relevant facts."
Source: http://www.stockholm-network.org

6.3.1 The Road to Serfdom

The Road to Serfdom by Friedrich Hayek was published in 1944. It immediately had tremendous success and was said to have been Margaret Thatcher's bedside book. Although the author was an economist, the book is more a philosophical pamphlet than an economic study. It does not contain any statistical data nor does it refer specifically to any historical episode. It argues that economic planning necessarily leads to totalitarianism but it is not obvious whether the demonstration is aimed solely at the fascist or communist regimes or also (and mainly) at the United States under the New Deal. We will see in a moment that it is probably the second conjecture which is correct.[10] It turns out that this fairly abstract essay had a print run of several million copies. This is the first point we must try to understand. Regarding the context of this publication one can mention the following features (Cockett 1994, Hayek 1994, McInnes 1998).

---

[10] Under the first assumption the usefulness of the book would have been limited for the fascist regimes were crumbling and the communist regimes were out of reach.

- Hayek was Austrian by birth but came to live in England in the mid 1930s. The manuscript of *The Road to Serfdom* was written in English but was translated into German and Spanish even before being published. As shown in Table 6.4 these translations were published in 1943, that is to say one year before the book was published in the UK and in the United States. Other translations followed closely.
- The book earned the unusual distinction of being reviewed at length in every national British newspaper. It was reviewed in a newspaper published in Christchurch, New Zealand, as early as May 1944, that is to say two months after its publication.
- *The Road to Serfdom* was reviewed twice in the *New York Times*, the first time on April 1, 1944 (p. 17) by Orville Prescott and the second time on September 20, 1944 (p. 21) by Henry Hazlitt. The two reviews are very different. Prescott speaks of a "sad and angry little book" written "with a fine contempt of easy readability". The *New York Times*'s negative posture was short-lived however (see Fig. 6.2a). Hazlitt[11] begins his review by saying, "In *The Road to Serfdom*, Friedrich Hayek has written one of the most important books of our generation," a fairly strong statement but one which indeed became true 35 years later.
- In April 1945 the *Reader's Digest* devoted the first 20 pages of its monthly issue to a condensation of *The Road to Serfdom*. As its circulation was about 8 million, Hayek became overnight a well-known figure in America. Moreover, the Book-of-the-Month Club distributed 600,000 copies of this condensation.
- In April 1945, *Look* magazine produced a cartoon summary of the book's main thesis, namely that planning leads to dictatorship. The cartoon was republished by General Motors later in 1945.
- In Spring 2005, Hayek went to the United States to give lectures in universities, but on arrival in New York he learned that the conferences were scheduled in city halls holding audiences of several thousand people. During his stay he also took part in several radio-broadcast round-tables.
- Yet, as a clear sign that Keynesian ideas were still solidly entrenched, Hayek did not easily find a permanent academic position in the United States. In 1950 he was a visiting professor at the University of Arkansas; subsequently he became a member of the Committee of Social Thought at the University of Chicago[12] but did not enjoy a permanent faculty position.

Any author will easily understand that a promotion campaign of such magnitude does not occur by pure luck. To get a better understanding of what happened one has to take a broader view.

---

[11] Hazlitt's support for Hayek's anti-Keynesian thesis does not come as a surprise. Through many of his articles in the *New York Times* he had commended books and ideas which called Keynes's ideas into question. As examples one can mention his review (January 1938) of Ludwig von Mises's book *Socialism, an Economic and Sociological Analysis*, or his review (October 1939) of Lionel Robbins's book *The Economic Causes of Class Conflict*. Both von Mises and Robbins were good friends of Hayek. In 1959, Hazlitt published a detailed chapter-by-chapter critique of Keynes's *General Theory*.

[12] The Committee was sponsored by the Volker Fund.

Table 6.4 *Promotional campaign of Hayek's book The Road to Serfdom*

Publication year	Country	Comment, publisher
1943	Spain	Spanish translation, University of Cordoba
1943	Switzerland	German translation, Reutsch
1944	England	English edition, Routledge
1944	United States	American edition, University of Chicago Press
1944	Sweden	Swedish translation
1945	France	French translation, Librairie de Médicis
1945	United States	Hayek's conferences, radio round-tables
1945	United States	Condensed version, *Reader's Digest*, Book-of-the-Month Club
1945	United States	Cartoon version, *Look* magazine
1945	United States	Cartoon version, General Motors (Thought Starter No. 118)
1945	Germany	Hayek's conferences in the British and US zones
1946	Denmark	Danish translation, Gyldendal
1946	France	Condensed version in French, Lhoste-Lachaume
1946	France	Third French edition, Editions Politiques, Economiques et Sociales
1948	Italy	Italian translation, published in Milan

Notes: The publication of 1943 in Spanish comprised only a few chapters of the book. The cartoon version (currently available on the Internet) is a crude representation which bears little resemblance to Hayek's book.
Sources: Online catalog of Harvard Library, British Library of Political and Economic Sciences (LSE), Digital Library of the Hochschulbibliothekzentrum (HBZ) of the Land of North Rhine-Westphalia.

6.3.2 Influence of business associations

The National Association of Manufacturers (NAM) claims to be the largest and most influential trade association in the United States. For the sake of brevity we concentrate our attention on this association but one should keep in mind that there are many other similar organizations. The NAM was created in 1896. For over a century, operating mostly from behind-the-scenes, it has been the most powerful organization representing the interests of large corporations. Between 1900 and 1980, the NAM was mentioned in 5,299 articles published in the *New York Times*.[13] In recent decades, due to the growing importance of non-manufacturing sectors, its influence has dwindled, but it continues to significantly influence US lawmakers.

---

[13] Broken up by decades the numbers are as follows: 1900–1909: 174; 1910–1929: 193; 1920–1929: 251; 1930–1939: 857; *1940–1949: 1908*; 1950–1959: 958; 1960–1969: 556; 1970–1979: 360. These figures clearly show that the decade 1940–1949 marked a peak in the activity and visibility of the NAM. Between 1935 and 1941, the membership of the NAM expanded more than threefold from 2,500 to 8,000 companies. Its public relations budget doubled from $500,000 to $1 million (Ewen 1996).

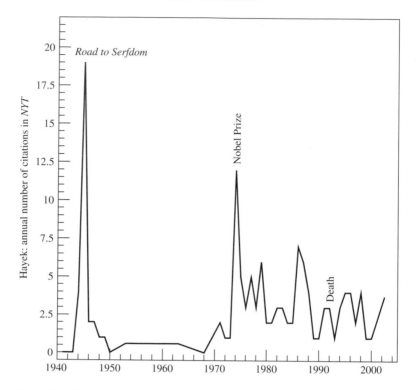

Fig. 6.2a Number of articles per year published in the *New York Times* which mention Friedrich Hayek. It is remarkable that the maximum occurred not after the Nobel Prize award but after the publication of *The Road to Serfdom* at a time when Hayek was relatively unknown. *Source: Website of the New York Times.*

Figure 6.2b provides a historical summary of the main themes on the NAM's agenda. Between 1934 and 1950 the main target was the New Deal policy of President Franklin Roosevelt. From 1950 until the late 1970s the crusade against Communism eclipsed all other themes; after that date, communism was less seen as a threat and the opposition to government "interference" again stepped into the foreground. To substantiate these statements we now briefly describe each of these periods.

Opposition to the New Deal The main reforms of Roosevelt's New Deal policy were approved by Congress in rapid succession during the first months of the new administration. With the Dow Jones index down by 85% with respect to 1929, this was a time of urgency and new measures had to be taken quickly. The National Industrial Recovery Act (NIRA) gave the president unprecedented powers over the economy and over business. The National Recovery Administration (NRA) created by the NIRA established minimum wages and maximum labor hours. Symbolized by a blue eagle, it was very popular with workers. Although

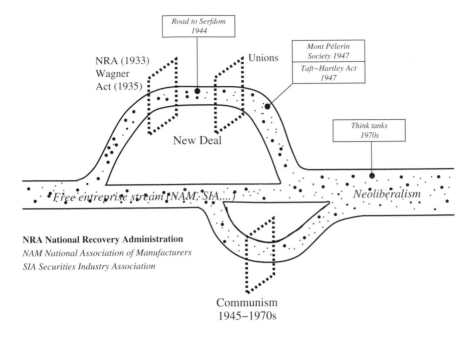

Fig. 6.2b Diagrammatic representation of public relations campaigns run by business associations, 1930–2000. The left-hand side of the stream corresponds to the early 1930s while the right-hand side refers to the late 1990s. The campaign against communism which took place between 1950 and the late 1970s marked a diversion from the mainstream objectives. Additional explanations about the Wagner Act, the Taft–Hartley Act, *The Road to Serfdom*, the Mont Pélerin Society and the neoliberal think tanks can be found in the text.

membership of the NRA was voluntary, public pressure made it almost mandatory. At first, the NAM urged its members to make the reforms succeed for the good of the country, but within six months its attitude changed. In 1935 the NRA was declared unconstitutional by the US Supreme Court. The president and Congress reacted by passing a number of acts whose purpose was to replace those invalidated by the Supreme Court. By the end of 1934 the *Times* summarized the situation in the following way.

Organized industry has declared open war on the New Deal and has announced its determination to campaign by every means in its power for the defeat of "President Roosevelt's new economic order". (*NYT* December 9, 1934, p. 23)

The objections of the NAM were mainly directed against the Wagner Labor Disputes Bill and the Temporary National Economic Committee (TNEC). For instance, in 1935 the NAM was trying to obtain a ban on general strikes and sympathetic strikes, that is to say strikes supporting the strikers at another company (*NYT* January 15, 1935 p. 3). One of the main activities of the TNEC was to enforce anti-trust legislation and in 1942 the NAM published a massive refutation (830 pages) of the arguments of the TNEC (Scoville and Sargent 1942). In mid 1943, while the war was far from being won, NAM officials called for a government commitment to a return to free enterprise in the post-war period (*NYT* June 22, 1943 p. 27). At the end of 1944, the NAM adopted a six-point program whose themes were basically those of *The Road to Serfdom* (*NYT* December 8, 1944 p. 1). The message against government intervention was repeated relentlessly in the late 1940s. Between 1946 and 1950 the NAM distributed 18 million pamphlets that pushed anti-New Deal, anti-union and anti-communist sentiments. A cartoon service serving more than 3,000 weeklies disseminated cartoons that humorized the NAM theses. For instance, the forgotten man that Roosevelt popularized in his speeches is represented as a tattered taxpayer. The NAM also produced radio programs (e.g. The Family Robinson) and movies, and it ran vast national billboard campaigns (Ewen 1996). These campaigns were indeed effective. On June 23, 1947, The United States Senate followed the House of Representatives in overriding Truman's veto and establishing the Taft–Hartley Act as a law. It amended the Wagner Act (which Congress had passed in 1935) in a way which was favorable to business. Similarly, while the prospect of universal, federally insured health coverage seemed close, the project was killed in Congress in November 1949.[14]

By mid 1950 with the beginning of the Korean War it was Communism which became the main target of the NAM. But before turning to this point we must describe what can be considered as a legacy of the struggle waged against the New Deal by the NAM. Joseph P. Kennedy, the father of President Kennedy, was a staunch supporter of President Roosevelt. He donated substantial amounts of money for Roosevelt's presidential campaigns in 1932, 1936 and 1940. President Roosevelt rewarded his support by appointing him first as chairman of the newly created Securities and Exchange Commission, and in 1938 as ambassador to the United Kingdom. It is clear that such a stand put him on a collision course with the NAM. As a matter of fact, this hostility seems to have been transmitted to Joseph Kennedy's sons. It is clearly apparent in an article that Senator John F. Kennedy wrote in the *New York Times* of February 19, 1956 (Magazine Section, p. 11). The article is entitled "To keep the lobbyists within bounds" and its first story is the denunciation of the bribery through which the NAM was "able to control the appointment of members to certain congressional committees." In 1958 the Kennedy–Ives labor reform bill was opposed by the NAM and defeated in the House of Representatives; another attempt made in 1959 met the same fate. This tense relation continued during Kennedy's presidency. For instance,

---

[14] A new attempt to pass a law in favor of universal health coverage was made in 1994 by the Clinton administration but was also defeated.

in May 1961 the *New York Times* writes, "The NAM found nothing right with President Kennedy's tax program"; it also opposed government outlays to combat the recession or the establishment of a system of health insurance for the aged. In December 1962 the *New York Times* notes, "President Kennedy had an antibusiness reputation before he even had the keys to the White House" (*NYT* December 17, 1962, p. 12). In May 1962 Republicans charged that President Kennedy had persecuted business to the point of provoking a stock market decline. Robert Kennedy was also the target of harsh criticisms by NAM officials (see for instance *NYT* April 8, 1961, p. 8).

The struggle against communism As this facet of the action of the NAM is somewhat outside of our main focus we will give only brief indications. As early as 1946 the US Chamber of Commerce distributed a million copies of a 50-page article entitled "Communism in the United States" and in 1947 there was a similar distribution of a pamphlet entitled "Communism within the Government" which alleged that about 400 communists held important positions in the government. In June 1947, the NAM announced that it had received demands for help from businessmen in Germany and Japan who were concerned about growing communist pressure and that it was planning to send aid abroad (*NYT* June 1, 1947, p. 50).

What were the connections between the NAM and critics of the New Deal such as Hayek and von Mises? Von Mises was being supported by the Volker Fund (see below) and was an economic adviser to the NAM throughout the 1950s (website of the Von Mises Institute). The connection between Hayek and the NAM is less clear but it is known that he delivered an address to the 66th Congress of the NAM (December 6, 1961). However, the main support came from business foundations: the John Olin Foundation, the Relm and Earhart Foundation, the Lilly Endowment and above all the Volker Fund. This is the topic that we discuss in the next section.

The Mont Pélerin Society: a nursery of Nobel Prize laureates It is the William Volker Charities Fund (established by the William Volker Company of Kansas City) which provided funding for von Mises' salary at New York University and for Hayek at the Committee of Social Thought of the University of Chicago. Harold W. Lubnow, the president of the Volker Fund, was a great admirer of *The Road to Serfdom* to the point that he offered Hayek $30,000 to write an American version of the book. It is likely that the Volker Fund provided financial support for the promotion campaign of *The Road to Serfdom*. As a matter of fact, the promotion of select books was one of the main activities of the fund. It was on the basis of reports established by two readers, Murray N. Rothbard and Rose W. Lane, that the fund selected the books which would benefit from extensive marketing campaigns.

Table 6.5 *Presidents of the Mont Pélerin Society who won Nobel Prizes*

Name	Term as president	Nobel Prize
F. Hayek	1947–1961	1974
M. Friedman	1970–1972	1976
G. Stigler	1976–1978	1982
J. Buchanan	1984–1986	1986
G. Becker	1990–1992	1992

Notes: The French economist Maurice Allais (Nobel Prize in 1988) attended the first meeting but did not serve as president. Three other prominent members of the Mont Pélerin Society were Ronald Coase, Vernon Smith and Erik Lundberg. Coase and Smith became Nobel laureates in 1991 and 2002 respectively; Lundberg was President of the Swedish Bank, a member of the Nobel Committee for the Prize in Economic Science from 1969 to 1979 and chairman of this committee from 1975 to 1979.

It is also thanks to the support of the Volker Fund that Hayek could organize the Mont Pélerin Society. The first annual conference of this society took place in April 1947 in Switzerland and brought together 39 people. It is of interest to take a closer look at this audience.

- There were four journalists, for the *Reader's Digest*, *New York Times* (namely Henry Hazlitt), *Fortune* and *Time and Tide* (London).
- Among the economists there were four future Nobel Prize winners: Maurice Allais, Milton Friedman, Friedrich Hayek and George Stigler.
- There were three economists from the Institut Universitaire des Hautes Etudes Internationales in Geneva (mentioned in Fig. 6.2c) and three participants were from the newly created Foundation for Economic Education.
- H. C. Cornuelle took part in the conference as a representative of the Volker Fund.

Of the 23 economists who served as president of the Mont Pélerin Society between 1947 and 2004, 5 became Nobel Prize winners shortly after the end of their terms as president, as can be seen from the Table 6.5.

The Nobel Committee comprises five Swedish economists (including its chairman) plus its secretary who is also an economist. From 1969 to 1979 one of its most influential members was Erik Lundberg. In 1979 he was replaced by Assar Lindbeck who chaired the committee from 1980 to 1994. With several other distinguished economists (among whom was Milton Friedman) Lindbeck contributed to the definition of an index of economic freedom, an idea which had been put forward at the 1984 biannual conference of the Mont Pélerin Society (Grubel 1998, p. 288). In 1994, Lindbeck coauthored a book with Torsten

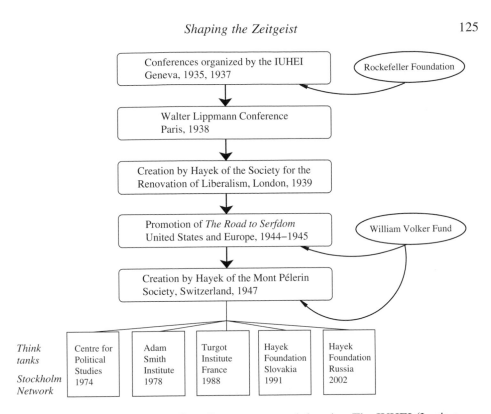

Fig. 6.2c Promotion of neoliberalism over several decades. The IUHEI (Institut Universitaire des Hautes Etudes Internationales) was established in Geneva in 1927 thanks to funding from the Rockefeller Foundation. These conferences were different from standard scholarly meetings in the sense that they had a public relations objective as revealed by the number of invited journalists. At the Mont Pélerin meeting of 1947 more than 10% of the 39 participants were journalists. Walter Lippmann who was himself a very influential journalist was a member of the Society for the Renovation of Liberalism which paved the way for the Mont Pélerin Society. *Sources: Cockett (1994), Coleman (1989), Cutlip (1994), Hayek (1994)*

Persson, another member of the Nobel Committee,[15] entitled *Turning Sweden Around*, which called for drastic cutbacks in Sweden's social security expenses.

What is the connection between the composition of the Nobel Committee and the orientation of the economics profession? A parallel with the US Supreme Court can be of some help. Like the Nobel Committee, the Supreme Court has only a small number of members: it consists of the Chief Justice and eight Associate Justices. Another similitude is the fact that their decisions are final and cannot be tested in another jurisdiction or committee. It is obvious that the composition of the US Supreme Court has a definite influence on its decisions,

---

[15] Persson became chairman of the Nobel Committee in 2002.

especially in so far as many decisions are reached with small margins. These judgments in turn shape the social and political climate in the United States. As its deliberations are not made public we do not know to what extent the choices of the Nobel Committee are influenced by its composition. However, there can be little doubt that the awards have a strong influence on the economics discipline. The Nobel Prize lends a powerful platform to the recipients and brings entire fields into the research spotlight. This effect is particularly marked in a time when normative research has pre-eminence over empirical and objective analysis, a distinction emphasized by Edmond Malinvaud (1990). It is unfortunate but probably revealing that the Nobel awards to Kuznets (1971), Leontieff (1973) and Stone (1984), which were meant to reward and encourage the vital but unglamorous tasks of data collection and economic measurements, did not bring about a perceptible rise of interest.

The creation of think tanks was one of the main modes of action of the Mont Pélerin Society. Pascal Salin, who served as president of the Society from 1994 to 1996, observed that more than one hundred neoliberal think tanks were created by members of the society throughout the world. As a matter of fact, several of the think tanks which belong to the Stockholm network are named after Friedrich Hayek: the Hayek Institute of Austria, the Friedrich Hayek Gesellschaft in Germany, the Hayek Foundation in Russia, the Hayek Foundation in Slovakia, the Hayek Society in Hungary, the Institut Hayek in Brussels.

Is there a link between these think tanks and the NAM apart from the fact that they advocate the same policies? This is fairly difficult to know for, in contrast to most political parties, think tanks do not have to disclose the origin of their funding. However, the fact that social events organized by think tanks are often held in luxurious hotels seems to suggest that there are generous sponsors.[16]

6.4 Tangible effects of neoliberal policies

What have been the tangible effects of neoliberal policies on the world economy over the past thirty years? This question is somewhat outside the scope of this chapter but it can hardly be avoided. It immediately raises another question: what is the best point of observation to answer this question? Europe may not be the best testing ground for, if one excepts Britain, the shift to neoliberal policies began only in the early 1990s. The United States may not be a good testing ground either, because as the dominant power this country is in the best position to take advantage of the globalization process.[17] The best point of observation

---

[16] For instance, the Workshop of European Think Tanks organized by the Stockholm Network in February 2005 was held in an up-market hotel in Brussels.

[17] In the words of a Peruvian political leader, "Some countries globalize and others are globalized."

seems to be Latin America. Why? (i) It is in Latin America that neoliberal policies were first implemented. After his accession to power in 1973, General Pinochet endorsed the program drawn up by Milton Friedman in his letter to him of April 21, 1975. This made Chile the first laboratory in which the economists of the Chicago school (the so-called Chicago boys) began to implement their policies.[18] Taxes were reduced for companies and wealthy households, unions were subdued and suppressed, state companies and social security were privatized, education was decentralized and partially privatized, and unemployment was maintained at a level which guaranteed that demands for higher wages would be under control. The fact that these reforms took place six years before Margaret Thatcher became Prime Minister in Britain emphasizes the pioneering role of the Chilean experience. Three years after Chile, Argentina followed suit. In subsequent years the neoliberal agenda was implemented in many Latin American countries: Bolivia and Mexico from the early 1980s, Peru under President Fujimori, Argentina under President Carlos Menem, Brazil under Presidents Collor de Mello and Cardoso.[19] (ii) In a sense Latin America was an ideal testing ground for privatization policies. Indeed, over previous decades many Latin American countries had developed an extensive state-owned sector. Mexico is a case in point: in 1984 the state controlled 1,212 firms and entities (Hart-Landsberg 2002). Such countries were therefore ideal for demonstrating the virtues of privatization policies. (iii) It is often recognized by lucid neoliberals that one of the side effects of neoliberal policies is to increase economic inequality. In several countries in Latin America (e.g. Brazil or Venezuela) income and wealth inequalities were already high in the early 1980s. This was likely to make this side effect less painful than in more egalitarian countries.

It would be futile to discuss the economic achievements of neoliberalism in Latin America. On the one hand, it can be observed that neoliberal mandates produced lackluster growth at best and often resulted in bankruptcies as in Mexico (1994), and Brazil (1999, 2002) or economic collapse as in Argentina (2000). On the other hand, these failures can be easily brushed aside by arguing that

---

[18] There had been close academic ties and exchanges between the economics department of the University of Chicago and the Catholic University of Santiago in the decade before the coup. After the coup, it was Chicago Professor Arnold Harberger who, through his frequent trips to Chile, served as go-between for the Chilean government and the Chicago School. Pinochet's economic options were endorsed by Milton Friedman in a letter exchange that followed their meeting in Santiago on March 21, 1975. Hayek's role was more indirect: in 1980 he became honorary president of the *Centro de Estudios Publicos* (CEP), a think tank whose board comprised Chilean rulers, heads of Chilean banking institutions and Chilean as well as American academics, e.g. Professor Theodore W. Schultz. Hayek made a clear distinction between totalitarianism and what he called authoritarianism. Thus, in a letter to the *Times* (August 3, 1978) he wrote, "I have not been able to find a single person even in much-maligned Chile who did not agree that personal freedom was much greater under Pinochet than it had been under Allende." (Walpen and Plehwe 2001).

[19] Incidentally, it can be noted that by observing the effects of neoliberalism in the region where it has been implemented with greatest strength we follow the methodological guideline introduced in a previous chapter as the extreme value technique.

corruption, cronyism and mismanagement had made the effective implementation of neoliberal policy almost impossible.

A more interesting question is whether neoliberalism was able to win a broad consensus, because this could be tested in free elections which took place in several countries. The net result was that neoliberal policies were flatly rejected by a majority of the voters in Venezuela (1998, 2000), Brazil (2003), Argentina (2003), Uruguay (2005), Chile (2006), Bolivia (2006) and Peru (2006). One could of course argue that the voters were misled by demagogues; but making people happy against their own will is a slippery road that many neoliberals would probably hesitate to advocate.

The previous judgment in terms of social consensus sheds light on only one of the facets of the question. Another important point concerns the achievements of neoliberal policies in industrialized countries such as the United States and the UK for which cronyism cannot be invoked to rationalize poor results. An assessment in terms of GDP growth rates is fairly tricky because growth is dependent on many exogenous factors that cannot be controlled. In contrast, an evaluation in terms of income inequality may be of greater significance because inequality is a structural factor whose evolution is not much affected by business fluctuations. Figure 6.3a shows that a notable increase in the share earned by the top 0.1% began to build up in the mid 1970s. This should be no surprise. Schematically, the allocation of income is the result of a balance of power between wage earners and employers. By defining the rules of the game, the state plays a crucial role in this competition. From a dynamic perspective the situation is somewhat similar to the equilibrium of a chemical reaction. As we know, the equilibrium is displaced when one of the factors (e.g. temperature or concentration of one of the reactants) is modified. Similarly, income allocation will be modified depending on whether the state sides with employees, as was more or less the case during the New Deal, or with employers, as has increasingly been the case since 1950. In fact, what is surprising is that the share of the top 0.1% did not begin to pick up earlier. This may be due to the momentum acquired between 1935 and 1950, a period in which the share of the top 0.1% fell steadily from 6% to 3.2%, just as the momentum of a ship keeps her on course for a while after the engine has been stopped. According to Fig. 6.3a the UK followed the same track as the United States with a time lag of about 10 years.

Neoliberal economists point out that an increase in income inequality is a natural consequence of a vibrant economy based on a free market. After all, does the experience of socialist countries not clearly show that egalitarianism leads to economic stagnation? If, pulled along in the wake of the wealthiest, the whole society experiences a steady improvement in welfare, education and other aspects, there can indeed be little objection to this argument. However, as

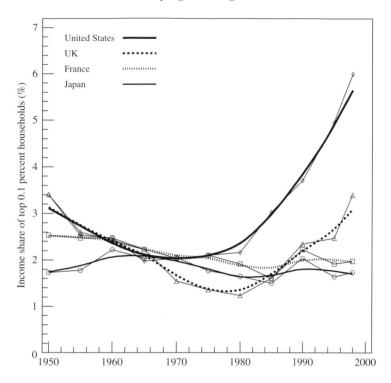

Fig. 6.3a Income inequality in several countries. Vertical scale: share of national income earned by the 0.1% of households with highest income. The data are based on incomes reported to the fiscal administration; they represent incomes before the payment of income taxes and exclude capital gains. Under an egalitarian distribution of income the top 0.1% would earn 0.1% of national income; in the United States their share is in fact 20 and 60 times larger in 1970 and 1998 respectively. *Sources: United States, UK, France: Piketty and Saez (2003); Japan: Moriguchi and Saez (2004).*

shown in Fig. 6.3b the reality is rather different. The real wage of non-managerial workers reached a peak level in the mid 1970s and has been on the decrease globally thereafter. Incidentally, one may wonder why this evolution differs so markedly from the trend of GDP per capita. The latter measure is a composite variable which, apart from wages, includes many other income components, for instance corporate profits or profits from the appreciation of stocks or real estate. The share of wages and salaries in national income was 65% in 1950, reached a maximum value of 76.4% in 1975 and then dropped steadily. In 2004 it was down to 45.4%; this represents a decrease rate of 1.07 percent per year. As a matter of comparison, the share of wages and salaries in the Australian national income was 54% in 2004.

Providing a good quality of education can be considered as another criterion which is fairly independent of business fluctuations. What is the picture in this

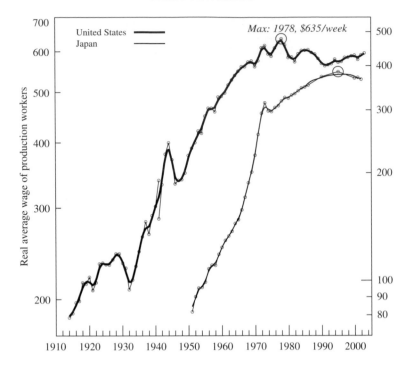

Fig. 6.3b Real wages in the United States and Japan. Thick solid line: real average earnings per week for production workers (manufacturing sector) in the United States expressed in 2000 dollars (left-hand scale); thin solid line: real average earnings per month for production workers (manufacturing sector) in Japan expressed in thousands of yen of 2000 (right-hand scale). Nominal wages were transformed into real wages by dividing them by the consumer price index. In 2003, US weekly earnings were 6.7% lower than in 1978. Earnings of manufacturing workers are the only data available on such a long time interval. For the broader category of production workers the evolution since 1950 is similar but there was in addition a reduction in the number of weekly hours in the period 1970–2003. *Sources United States 1914–41: Liesner (1989, pp. 98–9), 1941–2003: Website of the US Department of Labor; Japan 1951–86: Liesner (1989, pp. 266–7), 1987–2003: Japan Statistical Yearbook; see also Rodnesky (2004, p. 18).*

respect? The United States has a great number of splendid universities and in recent decades about 75% of the Nobel Prizes in physics, chemistry and physiology or medicine have been awarded to Americans. Yet surprisingly, after having increased for over 80 years at a fairly constant annual rate of about 5%, the number of Ph.D.s delivered annually (in proportion to total population expressed in millions) suddenly leveled off in the mid 1970s.[20]

---

[20] Between 1978 and 2000 the ratio remained at a level of about 13 Ph.D.s delivered annually per 100,000 population.

6.5 A network perspective

We focus on two aspects which both concern the relation between neoliberalism and democracy.

6.5.1 Implication for democracy

The notion of democracy plays a key role in our world but we often fail to realize that there are several bodies in our societies where democracy is absent, for instance enterprises, armies and churches. In his book *Corporation*, law professor Joel Bakan notes, "Deregulation is really a form of dedemocratization because it takes power away from a government elected by the people and gives it to corporations which are elected by nobody." As an illustration of this contention one can mention the fact that, even in the United States, government-sponsored companies such as the financial groups Fannie Mae and Freddie Mac are not only controlled by government agencies but also accountable to Congress. More than two centuries ago, Montesquieu already made the point that power without control sooner or later leads to autocracy.

6.5.2 Montesquieu and the balance of power

In his famous work *On the Spirit of Laws* (1748) Montesquieu observes, "Every man whom some power is granted tries to get more and eventually misuses it." It is this observation which led him to the notion that only a balance of power between different branches of government can ensure that none of them will take precedence over the others. Hayek wrote *The Road to Serfdom* in a time when states had concentrated in their hands enormous powers, rights and prerogatives. Quite naturally, he was concerned about the fact that this process might continue. However, instead of advocating a balance of power between states, corporations and employees he demanded almost unlimited power for business. Did he not realize that an atrophy of the state coupled with a collapse of unions would leave corporations a clear field? As a matter of fact, our present time seems to reproduce the characteristics of the society with which Montesquieu was familiar. In that time the contest opposed four players: the monarchy, the clergy (first estate), the aristocracy (second estate) and all other people who formed the so-called third estate. If one replaces "monarchy" by "central government", "clergy" by "media", and "aristocracy" by "corporate management" one gets a schematic picture of modern societies. Does this parallel provide a better understanding of the present situation? Of course it does not tell us what will happen but it provides possible scenarios.

- In some countries, such as Denmark or Sweden, the monarchy counterbalanced the power of the aristocracy, which produced fairly egalitarian societies and a smooth transition to democratic societies.
- In some countries which had a weak central state the aristocracy was almost unopposed; Poland was one of the best illustrations of this situation. As we know, the dominance of the nobility led to institutional paralysis, to the coming to power of puppet governments and eventually to the partition of the country.
- In a third class of cases, of which Russia is the best illustration, the central government was strong but, for historical reasons, consistently sided with the aristocracy. In economic matters Catherine II was an adept of the free play of market forces which however did not prevent her from pensioning off her lovers and courtiers with large estates and gifts of serfs. In the nineteenth century this led to a process referred to as the second serfdom. Even the abolition of serfdom in 1861 was more formal than real in the sense that peasants had to make redemption payments over a period of 49 years. Redemption payments were finally canceled in 1907, that is to say only three years before the end of the 49-year period.
- In France the situation was similar to the one in Russia in the sense that there was a strong central government which, especially after 1720, established a close alliance with the aristocracy. However, the third estate was much stronger than in Russia, which eventually led to the head-on confrontation of the Revolution of 1789.[21]

Which one of these roads will our societies follow in coming decades?

Appendix   From Edward Bernays to Isaac Asimov

Edward L. Bernays, who is considered the father of the public relations (PR) industry, wrote, "Those who manipulate the unseen mechanism of society constitute an invisible government which is the true ruling power of our country. We are governed, our minds molded, our tastes formed, our ideas suggested largely by men we have never heard of." Bernays sees a society in which opinion-molding tacticians are continuously at work, analyzing the social terrain and adjusting the mental scenery from which the public mind derives its opinion. There is a clear connection with the theme of Asimov's trilogy *Foundation* (1951, 1952, 1953)[22] but whereas Bernays' horizon was at most a few decades, Asimov's psychohistorians shape the history of our galaxy over several centuries.

Who was Edward Bernays (1891–1995)? During World War I, Bernays served on the US Committee on Public Information (CPI), the vast American propaganda apparatus mobilized in 1917 to package, advertise and sell the war

---

[21] Having a strong central state closely bound to the aristocracy, Britain belongs to the same class but the walls of the aristocratic class were more porous than in France which reduced the "osmotic pressure".

[22] Isaac Asimov (1920–92) is an American writer best known for his works of science fiction and for his popular science books.

as one that would "Make the world safe for democracy." The CPI would become the mold in which marketing strategies for subsequent wars, up to the present, would be shaped. Contrary to marketing techniques which are fairly well known, public relations techniques are used rather behind the scene. The ways and means through which it is possible to influence the Zeitgeist of a nation do not get much coverage. A meeting of scholars, the publication of a book, the writing of a newspaper article (or even of a "scientific" article) are hardly events which could be expected to change a society. It is only because these factors are used in conjunction with many others and over time spans of the order of several decades that they are effective. Their influence is gradual, almost unnoticeable. The purpose of this appendix is to give some indications about the techniques used in public relations campaigns and to place them in the context of a broader historical perspective.

The most fundamental technique of the PR industry is the so-called third party rule. It implies that the testimonial should come from people who do not seem to have any connection with the industry which pays for the campaign. It is mainly through this technique that PR distinguishes itself from marketing. An example can help to illustrate this point. The *Wall Street Journal* of July 19, 2004 published a full-page article about the possible risks of using cell phones while driving. However, the article gives none of the results which are mentioned in Table 6.1. How were the authors able to give a substantial account without mentioning these facts? From the public relations perspective the real challenge of such an article is to give the impression that the issue is taken seriously, yet without mentioning hard facts which would undermine the intended message. The vocabulary which is used provides a first indication: the word "safety" appears 15 times, whereas the word "accident" appears only once and the words "deaths" and "fatalities" do not appear at all. A common technique is to give the impression that it is a complex problem for which there is no clear solution.

It is at this point that recourse to the third party technique becomes essential. The article mentions research done at the University of North Carolina and published in the summer of 2003, which is quoted as saying "cell phones rank next to last on a list of common distractions for drivers." How can one understand that a study performed by university researchers can lead to results that are so obviously in contradiction with those summarized in Table 6.1? There are two explanations. (i) The study was funded by the American Automobile Association which, on this matter, is hardly an impartial player in the sense that its local sections sell cell phone services. Even if we assume that the funding did not influence the results of the study it can nevertheless influence the phrasing of the conclusions. As far as the general public is concerned it is the broad conclusion which matters. (ii) The complete release of the study contained reservations stating that the study had a number of limitations. Understandably these fairly technical qualifications were not reported in the account of the study released by the American Automobile Association which was quoted in the *Wall Street Journal*.

Another illustration can be found on the website of the Institut Turgot, a French think tank which is a member of the Stockholm network.[23] One reads (my translation): "The Turgot Institute hires and pays researchers for delivering scholarly studies showing the failure of state interventionism and the merits of an economic organization based on individual freedom and personal responsibility."

Although probably effective the previous techniques are relatively benign. In a book that Sheldon Rampton and John Stauber (2001) devoted to the methods of the public relations industry, the authors point out that more questionable methods are used occasionally to silence whistle blowers and spoilsport scientists. The threat of facing lengthy lawsuits is a very effective deterrent. Sometimes the threat becomes real, as in 1996 when the famous TV talkshow host Oprah Winfrey was sued over remarks made by one of her guests regarding the dangers of mad cow disease. Even though the law suit was finally dismissed in early 2000, its cost ran to millions of dollars in attorney fees. Many PR firms maintain extensive files on ecologists, public interest groups or scientists. There is of course nothing illegal about that but on occasion they use the information to discredit their opponents; needless to say, they do not release the information directly but resort to third parties. At this point the borderline between the methods used by intelligence agencies and those used by PR firms becomes rather fuzzy.

From the previous considerations one may get the impression that the PR industry is of relatively recent creation. This is certainly true as far as the industry itself is concerned. However, in previous centuries similar functions were carried out by other institutions. For an illustration let us come back to the episode of the Counter-Reformation that we already mentioned. As we know, the Jesuits were instrumental in stemming Protestantism in Bohemia, Hungary, Poland, Southern Germany, France and the Spanish Netherlands. Created in 1534 the new order was organized along military lines. They became preachers and confessors to people in power: kings, dukes, wealthy merchants. Throughout Europe they set up hundreds of schools and colleges attended by the sons of wealthy families, a method of opinion molding which was certainly as effective as the creation of think tanks. This was the peaceful side, but one cannot forget that the confrontation also led to the Thirty Years War (1618–48) which had a devastating effect on Germany.

---

[23] The founder and current president of the Institut is Jacques Raiman, a member of the Mont Pélerin Society, and its vice-president is Charles de Croisset, a vice-president of Goldman Sachs Europe.

7

Bonds of vassalage

Europeans celebrate the memories of German Chancellor Konrad Adenauer and of Robert Schuman, French Foreign Minister, for their role in European unification. However, President Harry Truman and Secretary of State Dean Acheson also played a key role and deserve great credit. If such a statement seems somewhat surprising nowadays it is only because the historical context of the 1940s has vanished from our memories. In response to the constitution of the Eastern Bloc under Soviet leadership, President Truman strongly backed the creation of a European Union as early as May 11, 1947 (*Le Monde* p. 1). In fact, the expression "European Union" appears in 567 of the articles published in the *New York Times* between January 1, 1945 and December 31, 1949. The expression "United States of Europe" appears in 182 articles; see below for the titles of three of these articles.

Year	Date	Page	Title
1945	Jan 1	8	Plan for Europe hailed: proposal for confederation gets support of 3 senators
1946	Nov 25	16	Winston Churchill goes ahead with his plan to form United States of Europe
1947	Apr 18	12	81 prominent Americans sign petition for United States of Europe

Moreover, Acheson's support was essential in the crucial move which led to the creation of the European Coal and Steel Community, often considered as the first step in European unification. Before delivering his famous declaration in the late afternoon of May 9, 1950, Schuman had consulted two people apart from his own government: Dean Acheson whom he met at the US embassy in Paris on May 8,[1]

---

[1] In a letter to Dean Acheson dated April 22, 1950, Jean Monnet who set up the project writes: "During the elaboration of the plan we have had several exchanges about its overall objectives." Thus, Acheson was already informed about the project when he met Schuman on May 8. On the contrary, Ernest Bevin, the British Foreign Secretary, was only officially informed on May 10, which aroused his anger (Roussel 1996).

and Konrad Adenauer to whom the project was submitted in the morning of May 9. The project was officially hailed by Acheson on May 11 and by President Truman on May 19 (*New York Times*, May 11, p. 1 and May 19, p. 3).

How many historical accounts of European construction mention the role played by the American diplomacy? An honest answer is that almost no account devotes more than a few lines to this question; see for instance the accounts given in Bitsch (1996) or Zorgbibe (1978, 2005) which reflect fairly well the mainstream literature. Even more telling is the omission made by Georges Bidault in his memoirs (Bidault 1965, p. 182). He was Prime Minister of the French government in May 1950 and gives a fairly detailed account of this historic and consequential episode, yet without acknowledging any contribution of the US diplomacy or any contact or meeting with the US Secretary of State on this matter.

Bidault's omission is typical. In a general way, political leaders and historians are reluctant to recognize that exogenous factors play a crucial role in the history of their countries. In the language of system theory this attitude can be summarized by saying that systems which are subject to many exogenous forces are in fact described as if they were closed. Naturally, this has disastrous effects on the soundness of the description. A physical parallel would be a pendulum which is swinging outdoors exposed to gusts of wind and whose trajectory one tries to explain without taking the influence of the wind into account. Nowadays, even the smallest countries are reputed to be fully independent and not in the least influenced by powerful neighbors. Saint Kitts and Nevis (42,000 inhabitants), a former British colony in the Caribbean which became independent in 1983, has a Prime Minister and a National Assembly of 14 members. Any allegation that the country might belong to the British or American sphere of influence would pass as unreasonable and misplaced.

It is important to realize that such a conception is relatively recent. In the nineteenth and early twentieth centuries the notion of zone of influence was a concept which was commonly acknowledged and used in diplomatic matters. For instance, a secret Anglo-French pact of 1916, the so-called Sykes–Picot agreement, put Syria in the French zone of influence and Iraq in the British zone. Subsequently, the League of Nations formalized this situation by giving France and Britain mandates over Syria and Iraq respectively. Nowadays, at least officially, there are no longer any mandates, protectorates, colonies or zones of influence. The fact that these expressions have been banned from our vocabulary made it difficult to find a title for this chapter. The expressions "satellite state", "vassal state" or "puppet state" have become so derogatory that it is hardly possible to use them. This is why we resorted to the medieval notion of vassalage. Lord and vassal were interlocked in a web of mutual rights and obligations;

whereas the lord owed his vassal protection, the vassal owed his lord military service and/or financial benefits. The relation between the United States and Australia is a case in point. In spite of the fact that it is the Queen of England who is Australia's head of state, the history of the country since World War II seems to suggest that the United States is its real lord.[2] In 1942 US troops protected Australia against a possible Japanese invasion; in return Australia has contributed troops in all major conflicts waged by the United States from Korea and Vietnam to Somalia, Afghanistan and Iraq.[3]

In this chapter we try to answer two questions.

(1) To what extent, in time of peace, can the history of a nation be affected by the interference and influence of powerful neighbors?

(2) How it is possible to detect and assess this influence despite the fact that the two parties usually prefer to keep it secret?

The second question calls for methods of investigation capable of seeing through standard historical accounts. In this chapter we propose and illustrate two methods. The first one relies on the identification of what we call *historical anomalies*, the second one is based on the observation of *coincidences*. This second approach parallels the coincidence method in particle physics in which one focuses on signals registered in coincidence in two (or several) counter tubes. As such coincidences can only come from two particles emitted in the same collision event, this method eliminates the influence of background noise and, by comparing the information provided by the two counters, allows identification of the event which produced the particles.

7.1 Role of the United States in the First Vietnam War

After 1945 the United States became the leading power of the western world. In a typical vassalage relationship, the United States covered 80% of the cost of the war that French troops were waging in Indochina against the Vietminh (*Quid* 1997, p. 1419c). Nobody would expect a country to cover almost all the costs of a conflict without having a say in strategic decisions and in the conduct of the war. This was even less likely in this case because of the overlord status of the United States, its superior warfare technology and the prestige of its generals.

---

[2] One should recall that in its medieval meaning the notion of vassalage refers to a hierarchy of bonds extending in descending degrees from the king (or emperor) down to the dukes, counts (or earls), barons and knights. Multilevel bonds may also exist in our present world; United States–Britain–Australia may be an example of a two-level system.

[3] Because its constitution does not allow Japan to send troops abroad, Japan contributed financially to some of these conflicts. Iraq was the first country to which Japanese troops have been sent since 1945.

Yet French historical accounts of the Indochina War contain almost no mention of the role played by the United States; see for instance the works by Lucien Bodard (1972–3, 5 volumes) or Georges Fleury (1994) which reflect fairly well the rest of the literature. It is this dichotomy and paradox which constitute what we call a *historical anomaly*.

First, let us recall that there were three main phases in the war.

- The period **1945–8** was marked by the end of World War II, the evacuation of the Japanese, the return of the French and lengthy negotiations between the French government and Ho Chi Minh, the leader of the Vietminh.
- **1949–52** The victory of the Communists in China was a watershed. All of a sudden the United States became greatly concerned about the spread of Communism in Indochina. Despite substantial military aid to the French troops the struggle remained uncertain however.
- **1953–4** In early 1953 Secretary of State John F. Dulles advocated a more offensive strategy. This led to the replacement of General Salan by General Navarre and to the adoption of the so-called Navarre Plan. Unfortunately, as recognized later in an article of the *New York Times* (May 16, 1954), this plan relied on faulty assumptions; it led to the defeat of Dien Bien Phu on May 8, 1954 and to the withdrawal of all French troops.

Chronology	The role of the United States in the First Indochina War

1945

| 1945, Sep. 26 | Lieutenant Colonel A. Peter Dewey, an OSS officer, was shot and killed and Captain Joseph Coolidge of New Hampshire was seriously wounded by Vietnamese during a revolt in Saigon against the return of French colonial rule. (*NYT* Sep. 27 p. 27 and Sep. 28 p. 1, Salisbury 1980, p. 38) |
| 1945, Sep. 29 | Admiral Lord Louis Mountbatten, chief of the Allied Southeast Asia Command, has sent reinforcements to Indo-China, where Vietnamese have been rioting for weeks. (*NYT* p. 5) |

Between 1946 and 1948 the French re-occupy Indochina; this turns out to be much more difficult in the north than in the south. American involvement remains very limited during this period.

After the Communist victory in the civil war in China, Indochina becomes an important strategic asset. As a result, the American Government becomes willing to back the French in their war against the Vietminh. At the same time, however, it does not wish to appear to support colonialism, which explains the contacts between American aid agencies and pro-independence, anti-Communist forces such as the Caodaists.

1949

| 1949, Jan. 25 | The French government has received assurance that Washington favors its efforts to check the Communist influence from the north that is expected to be intensified by the victory of the Communist forces in China. (*NYT* p. 18) |
| 1949, Apr. 1 | The French want US planes. (*NYT* p. 11) |

1950

1950, Feb. 28	France asks for US arms to fight Indo-China war. (*NYT* p. 20)
1950, Apr. 25	To facilitate the extension of military aid for the defense of Indo-China, US officials have asked the French government to draw up a plan for the future. (*NYT* p. 15)
1950, July 11	Indo-China awaits US military mission. (*NYT* p. 3)
1950, Oct. 5	General Brink will supervise US aid to Indo-China. (*NYT* p. 4)
1950, Oct.	France asked the US to furnish her a total of $3.1 billion in military assistance during the next year. (*NYT* p. 1)
1950, Nov. 5	US aid mission from the Economic Cooperation Administration (ECA) is hailed in Tay Ninh, home district of the Caodaist movement. Cheering and flag waving groups welcomed the US visitors (*NYT* 14 Oct, p. 10). The ECA is a board created in 1947 by President Truman to organize the economic aid provided by the Marshall plan. Cao Dai is a religion founded in Vietnam in 1926, which claims to combine the major religions of the world: Buddhism, Christianity, Confucianism, Hinduism, Islam and Spiritualism. In 1930, Cao Dai claimed 600,000 adherents; in 2005 it had an estimated 7 million adherents mainly in Vietnam. The Cao Dai Holy See is located in Tay Ninh, Southern Vietnam (*NYT* February 28 1930, p. 9 and Wikipedia, 2005). Cao Dai played an important part in the agitation for independence in the 1920s and 1930s. In June 1951, dissident Caodaist General Trinh Minh The broke away with a troop of about 1,000 supporters.
1950, Nov. 20	William C. Foster, Director of the Economic Cooperation Administration, avoids an ambush by Vietminh guerrillas outside Saigon (*NYT* p. 4).

Between 1951 and 1954, the war in Indochina gets progressively higher priority at the Pentagon and State Department. A continuous stream of high-ranking US officials visit Indochina in succession. In Saigon a number of US teams are in charge of dispatching the aid and advising the French military.

1951

1951, Jan. 21	Americans arrived in Hanoi to visit the front. (*NYT* p. 13)
1951, Feb. 21	Saigon: Communist-led Vietminh guerrillas fired mortars at the United States escort aircraft carrier *Windham Bay* in the Saigon River this morning and tossed six hand grenades into a bar crowded with United States sailors on shore leave from the ship this evening. (*NYT* p. 6)
1951, June 30	Preparation for signing Vietnam's first bilateral pact [between the United States and Indochina], namely an ECA agreement, were suddenly canceled because the procedure displeased Paris. (*NYT* p. 2)
1951, Aug. 1	A "human bomb" killed a French general, an Annamite provincial governor and himself with a grenade attack today in a crowded main street of Sadec, a village sixty miles south of Saigon. (*NYT* p. 1)
1951, Sep. 14	General de Lattre de Tassigny, French Far East commander, arrives in New York for a two week stay. He stated that a major purpose of his mission was to bring more information about the critical situation in Southeast Asia. (*NYT* Sep 14 p. 3, Sep 15 p. 14 and Sep 26 p. 2)
1951, Oct. 2	A shipload of United States Army Garand rifles, enough to equip four divisions [about 60,000 soldiers] was received in Saigon and turned over to Vietnam's army. (*NYT* p. 5)

1951, Oct. 23   General Collins, US Army Chief, is in Indochina to visit the front and confer on the war against the Vietminh (*NYT* p. 4). In October, there was also the visit of Congressman John F. Kennedy on a study tour of the Middle East and Asia. (Statement of Senator J. F. Kennedy, April 6, 1954)

1952

1952, Jan. 25   A dissident group of Caodaists, and not the Communist-led Vietminh forces [as announced in the January 10 edition of the *New York Times*] is held responsible for the latest outbreak of terrorism in Saigon. Twelve persons have been killed and more than eighty-five injured by delayed-action explosive charges set off on January 9, 1952 (*NYT* p. 3). [Subsequently, this attack was picked up by Graham Greene as the theme of his novel *A Quiet American* published in 1956.]

1952, Jan. 29   In the 16 months since August 1950, 100,000 tons of US military supplies have been delivered to forces fighting the Communist-led Vietminh insurgents. (*NYT* p. 3)

1952, June 25   Brigadier-General Francis G. Brink, Indo-China team chief, who returned to Washington two weeks ago, was found shot in an office in the Pentagon building and died in an ambulance en route to Walter Reed Hospital. (*NYT* p. 15)

1952, June 27   Ambassador Donald R. Heath leaves for Saigon post. (*NYT* p. 6)

An ambitious plan is set up in the spring of 1953 through which a decisive victory is expected. After the plan gets the approval of the Joint Chiefs of Staff, General Navarre is put in charge of its implementation, advised by a group of four American liaison officers.

The flow of American visitors intensifies. The new strategy of establishing strongholds in isolated places to lure and wear out Vietminh forces is at first successful as reflected in the optimistic comments of the New York Times.

After the set-back of Dien Bien Phu the government of Pierre Mendès-France takes the decision to end French involvement in Indochina and to negotiate an agreement with the Vietminh. In Saigon, the United States back Ngo Dinh Diem and, in January 1955, begin to train Vietnamese forces.

1953

1953, Mar. 1   John Gunther Dean arrived in Saigon as a financial adviser. In an interview given on September 6, 2000 he declared: "Few people realize today that the French Expeditionary Corps and the Vietnamese Armed Forces were all financed by the United States. My job was to document how the money was spent for example for pay, ammunition training of Vietnamese or Cambodian pilots, etc." (http://www.jimmycarterlibrary.org/library/oralhistory)

1953, Mar. 20   General Mark W. Clark, the US Commander in the Far East, arrived in Saigon for a four-day visit to Indo-China. (*NYT* p. 1)

1953, Mar. 27   The United States agreed to increase its contribution [to the Indochina war] but insisted that France in return should produce a program for winding up hostilities in a victory. (*NYT* p. 1)

1953, Apr. 3   General John W. O'Daniel visits Saigon. (*NYT* p. 5)

1953, Apr. 8   Mr. Adlai Stevenson, former US delegate to the United Nations, held a press conference in Saigon at the end of a week's visit to Indo-China. (*NYT* p. 5)

1953, Apr. 21   In a memorandum for the Secretary of Defense dated 21 April 1953, subject "Proposed French strategic plan for the successful conclusion of the war in Indochina", the Joint Chiefs of Staff pointed out certain weaknesses of the French plan but felt that it was workable. (*Pentagon Papers*, Volume 1, Document 17)

1953, Apr. 26   Admiral Radford, US Commander in the Pacific, arrived in Hanoi this evening. He is worried by the Vietminh invasion of Laos. Its capital, Luang Prabang, is supplied by the French Air Force which uses a number of transport planes on loan from the United States Far East Air Force. (*NYT* p. 4)

1953, May 9   Criticized on the war, the French Cabinet named General Henri Eugène Navarre as commander in chief in Indo-China in replacement of General Raoul Salan. General Navarre was previously Chief of Staff of the Allied ground forces in Central Europe under General Matthew B. Ridgway. (*NYT* p. 1)

1953, July 10   Lieutenant-General John W. O'Daniel, Commander of the US Army forces in the Pacific, said tonight as he wound up a three-week survey tour of Indo-China that he would recommend an increase of US military aid (*NYT* p. 2). During his visit, General Navarre submitted in writing to him a paper entitled "Principles for the conduct of the war in Indochina" which presents a marked improvement in French military thinking. Repeated invitations were extended to the US mission to return to witness the progress the French will have made. To improve the chances of success, [US] support should include close liaison with French military together with friendly but firm encouragement and advice where indicated. (*Pentagon Papers*, Volume 1, Document 17)

1953, July 13   John Foster Dulles, Secretary of State, expressed great satisfaction today with a new French military plan designed to regain the initiative in the offensive against Communist forces in Indo-China. (*NYT* p. 1)

1953, July 21   Highly successful French paratroop raid on the Communist supply base at Langson. (*NYT* p. 22)

1953, Aug. 28   US support should be conditioned upon French willingness to receive and act upon US military advice. Further, the French should be urged to vigorously prosecute the Navarre concept to the maximum extent of their capabilities. (Excerpts from a memorandum of the Joint Chiefs of Staff for the Secretary of Defense, *Pentagon Papers*, Volume 1, Document 17)

1953, Sep. 16   The Senate Foreign Relations Committee dispatched Senator Mansfield on an inquiry mission to Indo-China. (*NYT* p. 1)

1953, Oct. 27   The United States Legion of Merit was awarded today to a French woman Captain Valérie André, an army doctor and pilot, for her valiant services in Indo-China. (*NYT* p. 2)

1953, Dec. 19   French-Vietnamese forces set up a base in Dienbienphu to harass Communist flanks. (*NYT* p. 3)

1954

1954, Jan. 30   During a meeting of the President's Special Committee on Indochina, Allen W. Dulles [Head of the Central Intelligence Agency] inquired if CIA Colonel Ed Lansdale could not be added to the group of 5 liaison officers to which General Navarre had agreed. (http://www.totse.com/en/politics/central_intelligence_-agency/166660.html)

1954, Feb. 11   President Eisenhower asserted today that he could conceive of no greater tragedy than for the United States to become involved in an all-out war in Indo-China. (*NYT* p. 1)

1954, Feb. 16   The United States has asked René Pleven, French Minister of National Defense, to visit Washington after he completes his investigation of the Indo-China situation. (*NYT* p. 3)

1954, Feb. 18   President Eisenhower asserted that the United States was not trying to help anyone maintain colonialism in Indo-China. He repeated that the war there is a fight for the independence of the Vietnamese people. (*NYT* p. 1)

1954, Feb. 20   Vietminh's defeat in 1955 is predicted. Foe are checked on all fronts. (*NYT* p. 2)

1954, Mar. 7   The French destroyed key bases of Caodaists. (*NYT* p. 3)

1954, Mar. 8   US Information Agency unit spreads ideas in Vietnam. (*NYT* p. 11)

1954, Mar. 13   Lieutenant-General John W. O'Daniel will head the US military assistance group in Indo-China. (*NYT* p. 2)

1954, Mar. 23   Immediate dispatch of a new group of 25 US B-26 bombers to Indo-China to reinforce French Air Force. (*NYT* p. 1)

1954, Mar. 26   Fire bombs halt Vietminh attack. (*NYT* p. 2)

1954, Mar. 26   Admiral Radford proposed to General Ely the support of American bombers based in Manila: about 60 B-29 bombers would mount night raids dropping 450 tons of bombs each time. They would be escorted by 150 fighters from aircraft of the Seventh Fleet. Their objective would be to pulverize the ground round Dien Bien Phu from which the Vietminh were mounting their offensive. Paris gave its agreement to this scheme, but "Operation Vulture", as it was code-named, was vetoed by Congress on April 5. One month later, on April 23, Prime Minister Georges Bidault asked Mr. Dulles if the United States could not reconsider its decision and authorize the carrying out of "Operation Vulture"; the request was rejected by Mr. Dulles on the next day. (*NYT* January 20 1960, p. 10)

1954, Mar. 27   General Paul Ely, French Chief of Staff, flew back from Washington to Paris with a promise of 25 additional B-26 light bombers for Indo-China. (*NYT* p. 2)

1954, Apr. 6   "If the French persist in their refusal to grant the legitimate independence desired by the peoples of the Associated States [of Cambodia, Vietnam and Laos], it is my hope that Secretary Dulles will recognize the futility of channeling American men and machines into that hopeless internecine struggle." (Statement of Senator J. F. Kennedy in Congress)

1954, Apr. 8   The cornerstone of the US and French strategic policy in Indo-China for the last year has been the so-called Navarre plan. But Dulles' views are now considered to be too optimistic. (*NYT* p. 4)

1954, Apr. 16   Major-General John W. O'Daniel arrived in Saigon. (*NYT* p. 2)

1954, May 8   (a) Two US pilots die in Indo-China war. Their Flying Boxcar blew up yesterday on a supply drop mission to Dienbienphu killing its two US civilian pilots and the French crew chief. (*NYT* p. 2)

1954, May 8   (b) Dienbienphu is lost after 55 days. Dulles says that unity can check reds. France is sending more men to war. Shock of loss seems to unify deputies. (*NYT* p. 1)

1954, May 16   Lessons of Dienbienphu: US military policy was based on faulty intelligence. The French arrested intelligence agents working for the United States. (*NYT* p. E5)

1954, June 8	The Defense Department plans to replace man for man the 200 Air Force technicians it is recalling from Indochina. (*NYT* p. 1)
1954, June 7	The *Washington Post* and *Times-Herald* said today that twice during April the United States proposed using Navy and Air Force planes based in the Philippines to intervene in the Indochina war, provided Congress and allied nations agreed. (*NYT* p. 3)
1954, July 22	President Eisenhower reluctantly accepts the Indochina accord signed in Geneva. He asserts that the United States will not be bound by armistice terms. (*NYT* p. 1)
1954, Aug. 21	The United States expects the French to leave behind for the Vietnamese army the military equipment that it supplied for the Indochina war. (*NYT* p. 3)
1954, Oct. 9	Anti-Communist refugees are taken by the US Navy from Hanoi to South Vietnam. (*NYT* p. 3)
1954, Dec. 12	The United States will take over the training of South Vietnam's National Army about January 1, 1955. (*NYT* p. 4)

Notes: NYT means *New York Times.* The table provides a glimpse of US involvement based on American sources; in order to get a more comprehensive picture, Vietnamese sources would certainly be useful as well. Parallel episodes are (i) the aid brought by British troops to the Dutch in the warfare against independence movements in Indonesia (1945–7); (ii) the aid provided by the United States to Guomindang forces in the civil war against Communist forces (1945–9). These cases are interesting because they show a gradation in the forms of the support. In Indonesia there were 20,000 British (partly Indian) troops who took part in the fighting, in China there were 65,000 US troops who did not take part in the fighting. In Indochina there were no US troops (except for a few hundred airmen) but the United States covered about 80 percent of the expenses of the war. Many thanks to my colleagues Olivier Gérard and Dietrich Stauffer for their help in establishing this chronology.

In a sense this outcome was predictable because of fundamental opposition between French and American goals. While the French tried to restore and preserve the former colonial rule, the US government proclaimed repeatedly that it favored the full independence of Vietnam provided it was not under a Communist government (see the Chronology at the date of Feb. 18, 1954). This opposition was illustrated by two episodes: (i) US aid to nationalist movements such as the Caodaists (see the Chronology at the dates Nov. 5, 1950; Jan. 25, 1952); (ii) American attempts to make agreements with the Vietnamese government over the head of the French (see the Chronology on June 30, 1951). In 1954 there was another obstacle, namely the disagreement between Dulles's ambitious plan and President Eisenhower's intention to limit American involvement (see the Chronology at the dates of July 13, 1953 and Feb. 11, 1954).

Regarding the diverse aspects of the American involvement, the chronology suggests the following conclusions (the dates preceded by the letter "C" refer to the chronology.

US aid The aid consisted of supplies, air force technicians and pilots (C: May 8
(a) and June 8, 1954), pay of French and Vietnamese troops (C: Mar. 1, 1953),
training in psychological warfare (C: Mar. 8, 1954) and economic aid (C: Nov. 5
and 20, 1950).

Monitoring Between 1950 and 1954 there was an uninterrupted stream of
high-ranking American officers, senators and political personalities who visited
Indochina. For instance the US Army Chief (C: Oct. 23, 1951), the US Commander in the Pacific (C: Apr. 26, 1953), Senator Mansfield (C: Sep. 16, 1953),
General O'Daniel (C: Apr. 3, 1953, July 10, 1953, Apr. 16, 1954). In addition
there was a permanent US military assistance group headed successively by General Francis G. Brink, Ambassador Donald R. Heath and General John O'Daniel.
Often strategic decisions were made during conferences held in Washington with
French officers and political leaders (C: Feb. 16, 1954).

Recommendations The French Commander, General Navarre, had five American
advisers (C: Jan. 30, 1954). In addition he had to submit his plans in writing
to the Pentagon or to US officers who visited Indochina on inspection tours (C:
Apr. 21, 1953 and Jul. 10, 1953). American aid was continued only on approval
of the plans and was conditional on further supervision (C: Aug. 28, 1953).
We will probably never know to what extent American pressure was determi-
nate in the replacement of General Salan by General Navarre, but we know
that the plan which subsequently became known as the Navarre Plan was submitted to the Pentagon in April 1953, that is to say a few weeks before the
nomination of General Navarre as Commander in Chief in Indochina. It can
also be observed that in the first months after the "Navarre Plan" was put into
effect, the accounts of the *New York Times* became surprisingly optimistic (see C:
Feb. 20, 1954)

It is difficult to assess how much leeway was left to French officers. American
military planners are renowned for their attention to detail, which would imply
that US recommendations reached down to tactical level. If this assumption is
true, it provides a possible explanation of the overall failure; indeed from a
network science perspective the fact that the decision makers are located far
away from the war theater creates inherent liability and weakness; this effect
will be examined in a subsequent chapter under the name of *absentee ownership
syndrome*.

In the present episode it is fairly clear that continuation of financial and technical
aid was the main means of control; however, one may wonder what are more
generally the ways and means used by a country in order to influence another?
Some answers are provided in the next section.

7.2 Ways and means

In accordance with the notion of vassalage the most natural way to establish control over another country is by guaranteeing its security and defense. Naturally, the degree of dependence can vary greatly from case to case. After World War II, in the face of the threat (real or supposed) of a Soviet invasion, almost all West European countries were dependent on the United States for their defense. The following episode illustrates how this lever could be used. In January 1946, after the resignation of General de Gaulle, the formation of a Socialo-Communist coalition government was contemplated in France. During a crucial meeting an urgent letter was delivered by a motorcyclist. It was written by General Billotte, the Deputy Chief of Staff, and explained that "a Socialo-Communist government would be seen as a threat by our Allies; as a result they may consider reducing their commitment to guarantee our security" (Demory 1995). Eventually, a coalition government was formed which, apart from the Socialist and Communist parties, also included a Christian democratic party (MRP). Probably we will never know what was the real influence of General Billotte's letter. As echoed by the *New York Times*, a similar warning was given to French politicians in May 1958 during another political crisis: "United States military officials, deeply concerned over the developments in France, are reconsidering proposals to relocate important European military installations" (*NYT* May 19, 1958, p. 1).

A fairly discreet way of keeping a handle on public opinion is to impose some form of control on the media. We use the term control rather than the term censorship to emphasize that usually it is a double-effect control in the sense that some news items are amplified while others are restricted or suppressed. As illustrations of the amplification effect one can mention two episodes. (i) On March 5, 1946 a German editor at the *Neue Zeitung* received a phone call from General Eisenhower's headquarters asking him to devote the entire front page of the next edition to Churchill's Fulton speech (held in Fulton, Missouri, it popularized the phrase "Iron Curtain" to describe the division between Western powers and the area controlled by the Soviet Union and is considered as marking the beginning of the Cold War). He complied but left the *Neue Zeitung* on March 11, 1946. (ii) During the Cold War, German radio stations participated in anti-Communist campaigns in that they were required to grant airtime to the Allied Military Government for its broadcasts. An example of the censorship aspect is provided by the "Allied High Commission Law concerning the Freedom of Press, Radio, Information and Entertainment" which was issued in West Germany on September 22, 1949. Its first article says: "The German press and radio shall be free." Yet it is immediately corrected by Article II which stipulates, "any person engaged therein shall not act in a manner likely to affect prejudicially the prestige and security of Allied forces. Where in the opinion of the Allied High Commission a person has violated this provision, the Commission may prohibit the person from continuing its activities." Article III completes the control by providing that "no new radio broadcasting or television shall be set up without the authorization of the Allied High Commission" (Hartenian 1984, pp. 125, 126, 185).

The discreet presence of liaison officers or financial experts is a key technique for supervising another country. This can be illustrated by an example drawn from the British occupation of Iraq in the 1920s. Intelligence agent Gertrude Bell played a major role in setting up the Iraqi government under British supervision. In a letter to her father (Bell 1924), she explains that by the treaty between Britain and Iraq there are 18 reserved posts for British officials in Iraqi ministries and 5 posts as judges. Furthermore, apart from these reserved posts ministers had to put up lists of British advisers whom they considered necessary. Altogether there were 17 British advisers in the Ministry of the Interior and 15 in the Police Department. These advisers were appointed under long term contracts for periods varying from 5 to 15 years and most of them remained in their posts even after the country became formally independent in 1932. As this example shows, peace treaties are frequently used to impose constraints limiting the sovereignty of the dependent government. As another illustration one can mention the Cuban treaty of 1904 with the United States. Under this treaty it was impossible for Cuba to enter into any foreign alliance or to make broad changes in internal policy without the acquiescence of the United States. Often, especially for small countries, entering into a loan arrangement with a big power is the beginning of vassalage; first the loan may impose drastic guarantee conditions such as the fact that import and export duties may go to the lender if interest payments are delayed. In a second phase a financial commission headed by representatives of the lender may be appointed which has the power to control the whole financial sector (Nearing and Freeman 1926).

In many cases it is not possible to get direct information about moves which take place behind the scenes. In such cases one can nevertheless detect exogenous forces provided they affect several countries. The next section gives an illustration of this method.

7.3 Identification of interference through the coincidence method

Communism has been a concern for the US government at least since 1917 but the year 1950 marked a climax of anti-Communist activity. This is shown in a quantitative way in Fig. 7.1. The fact that the curve peaks in 1949 is quite understandable for it was in this year that the Communists came to power in

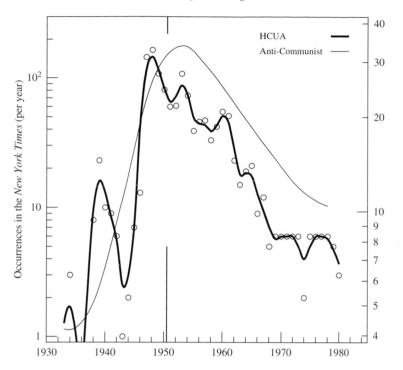

Fig. 7.1 Fluctuations of anti-Communism in the United States. Thick solid line and circles (left-hand scale): number of *New York Times* articles per year containing the two expressions "Communists" and "Committee on Un-American Activities"; thin solid line (right-hand scale): number of *New York Times* articles (in five-year intervals) containing the word "anti-Communist". The vertical lines signal the beginning of the Korean War. The House Committee on Un-American Activities (HCUA) was set up in 1933; its mandate was to get information on how foreign subversive propaganda enters the United States and about the organizations that are spreading it. *Source: Electronic index of the New York Times.*

China and that the Soviet Union successfully tested its first atomic bomb. Finally, after the outbreak of the Korean War in June 1950, the Cold War became a real war. In such circumstances broad scale initiatives and actions aimed at containing Soviet influence could be expected. We will see that this led to the creation of several anti-Communist organizations in the months after June 1950. Table 7.1 lists some of these episodes but it is by no means exhaustive. In fact, there were also similar episodes in Denmark, Switzerland, Greece and Turkey (Dubois 2003).[4] Even more than the near simultaneity of their creation, it is the fact that these movements were built on the same pattern that points to a

---

[4] Unfortunately, no detailed information is available for these cases.

Table 7.1 *Anti-Communist organizations that sprang up in the early 1950s*

	Country	Year	Month	Organization	Bulletin
1	West Berlin	1950	Jun	Congress for Cultural Freedom	
2	West Germany	1950	Aug	Volksbund für Frieden und Wahrheit	*Die Wahrheit*
3	Italy	1950	Aug	Atlantici d'Italia	
4	Belgium	1950	Aug	Paix et Liberté	
5	France	1950	Sep	Paix et Liberté	*Défendre la Vérité*
6	Netherlands	1951		Vrede en Vrijheid	*Die Echte Waarheid*
7	Australia	1951		Australian Ass. for Cultural Freedom	
8	Europe	1951	Aug	Comité Européen Paix et Liberté	
9	Italy	1953		Pace e Libertà	

Notes: In cases 1 and 2 funding by the Central Intelligence Agency is well established; in the other cases it can only be inferred from the similarities between the movements. The "Congress for Cultural Freedom" organization was founded on June 26, 1950; it sponsored about 20 publications in various countries, e.g. *Encounter* in the UK, *Preuves* in France, *Quadrant* in Australia, which were specifically aimed at intellectuals. In Belgium the organization "Paix et Liberté" was headed by Marcel de Roover, an industrialist who later took part in the creation of the "World Anti-Communist League" (Taipei 1967). *Sources*: Flamigni (2004), Delmas and Kessler (1999), Depraetere and Diericks (1986, pp. 83, 91, 244); Ludwig (2003); http://www.lurojansen.nl.

common origin. This can be illustrated by the cases of France, Germany and the Netherlands.

- **France** In September 1950, with the support of Prime Minister René Pleven, Jean-Paul David, Deputy of Seine-et-Oise, created the organization "Paix et Liberté" (Peace and Freedom) in order to counter the influence of the Communist party. During a visit to Washington in February 1952, Jean-Paul David met with Secretary of State Dean Acheson as well as with John Foster Dulles who was to succeed Acheson in 1953. Among the means and media that were used by the organization one can mention: (i) billboard campaigns: a total of 38 million color posters were printed over a period of time which goes approximately from 1950 to 1956; (ii) cartoon booklets; (iii) specific publications aimed at teenagers and women; (iv) an information bulletin entitled *Défendre la Vérité* (Supporting the Truth); (v) movies, one of them celebrated the French contingent in Korea.
- **Germany** The "Volksbund für Frieden und Freiheit" (the People's Union for Peace and Freedom) was created on August 29, 1950, that is to say, a few weeks before its French counterpart. It was funded by the United States but was also supported by the West German government. In March 1952, the VVF was granted the status of a state-approved organization. The propaganda means used by the VVF included posters, booklets, movies, and an information bulletin entitled *Die Wahrheit* (The Truth).

- **The Netherlands** The organization "Vrede en Vrijheid" (Peace and Freedom) was created in 1951 with official support from the government. Its bulletin entitled *De Echte Waarheid* (The Real Truth) contained cartoons, some of which were contributed by Fritz Behrendt, one of the most famous Dutch illustrators.

By our present standards, making use of color posters, cartoons and movies does not seem surprising or unusual. However, in the early 1950s very few European political parties were using such promotional means. In contrast, such modern public relations techniques were commonly used in the United States. As mentioned in an earlier chapter, the National Association of Manufacturers used cartoon services, vast billboard campaigns, radio programs and movies.

The synchronicity, the similarities in names, titles and methods are so obvious that one can hardly doubt that these organizations were set up on the same mold and had a common origin. This shows the power of comparative analysis; no similar conclusion could have been drawn from an analysis conducted at the level of only one country.

8

The absentee ownership syndrome

The central theme of this book is the analysis of interactions in social systems. As a first (and perhaps easier) step we will study the consequences of a *lack of interaction* between social agents. For instance, what happens when there are no interactions between landowners and tenants, holding companies and employees, governments and subjects? These questions might seem fairly difficult, but fortunately we have a good starting point because we can rely on the results of a famous experiment performed by Stanley Milgram in the 1960s.

First of all, to introduce the issue of absentee ownership, we describe landownership in the Philippines and Japan as seen by General Douglas MacArthur.

8.1 Land reform in Japan under General MacArthur

After General MacArthur retook Luzon in March 1945, some of his officers suggested that he send a punitive expedition against the Huks who were waging a guerrilla war in Central Luzon to dispossess the landlords. He refused and justified his position in the following way (Manchester 1978, p. 420).

Tarlac [located 100 kilometers to the north-west of Manila] marks the border between the sugar economy and the rice country. North of them the people grow rice and most of them own small areas of land. Did you notice how many schools there are up there, how the people dressed, looked happy? Do you see the hangdog look they have here, resentful, poorly dressed? Most of this land is owned in Madrid or Chicago or some other distant place. This is really *absentee ownership*. No pride, few schools, little participation in government. This is where organizations like the Hukbalahaps are born and get their strength. They tell me the Huks are socialistic, but I haven't got the heart to go after them. If I worked in those sugar fields I would probably be a Huk myself.

MacArthur had first-hand knowledge of the Philippines where he had spent 15 years in various positions. He was hardly a socialist, and at some points of his career he aligned himself with the right wing of the Republican Party. Yet, as

150

Supreme Commander of Japan he proved that his aversion to absentee ownership was not pure rhetoric. As a matter of fact, he initiated a sweeping land reform. Before that reform, power resided in a rural oligarchy of some 160,000 absentee landlords, each of whom owned on average 36 farms (Manchester 1978, p. 508). In December 1945, that is to say three months after the beginning of the occupation of Japan, MacArthur told the old Diet to pass a drastic land reform. However, the law which was passed by this assembly exempted 70% of the land from the reform. At MacArthur's insistence, a more effective land reform law was passed one year later by the first post-war Diet. All land held by absentee owners was subject to compulsory sale to the government or to the tenants. Because sale prices were set without taking into account the high inflation rate they were absurdly low, which made it easy for the tenants to become the new owners. Characteristically, MacArthur made the comment: "since the Gracchi effort of land reform in the days of the Roman Empire, there has been nothing quite so successful."[1] MacArthur's action was not limited to the farming sector. A total of 115 holding companies were dissolved. We can recall that a holding company does not produce goods or services by itself but controls or owns other companies by holding part of their stock or other financial assets.

The implications of the reforms initiated by MacArthur in terms of network connectivity and economic efficiency will be discussed later in this chapter. In the next section we describe the major discovery made by Stanley Milgram regarding the implications of an absence of interpersonal interactions.

8.2 How the strength of interpersonal interactions conditions human behavior

Stanley Milgram (1933–84) was one of the most influential psychosociologists of the twentieth century. Two of his experiments are particularly famous: the small world experiment and his experiments on obedience to authority. In this section we describe the second. Carried out in the early 1960s the obedience experiments raised great interest among the general public because they seemed to "explain" the obedience attitudes that had been observed in Nazi concentration camps. However, this interpretation belittles the significance of Milgram's experiments. Numerous historical episodes, from the repression of the Paris Commune in 1871 to the Katyn and My Lai massacres show that, under certain conditions (to be examined shortly), obedience to authority is a standard characteristic of human behavior. Reproducing this behavior in a laboratory experiment would

---

[1] In fact, the reform initiated by Tiberius Gracchus (162 to 133 BC) and Gaius Gracchus (154 to 121 BC) took place more than a century before the Roman Republic became an empire. To MacArthur's credit it should be added that history tells us that land reforms have more often foundered than succeeded.

add little to our understanding. As a matter of fact, the purpose and significance of Milgram's experiment are much deeper. He has shown that the weaker the interaction is between two persons, the easier it is for one to harm the other. Through his experiments Milgram was able to give this proposition a precise quantitative meaning. To understand how he formulated this result we need to know more about the experimental procedure.

The experiments involved three individuals (Fig. 8.1):

- the experimenter E who was also the supervisor;
- the instructor I;
- the subject S.

Both E and S were members of the experiment team whereas I was recruited through a newspaper advertisement and was paid $4.50 for one hours work. At the beginning E explains to I that the experiment is a scientific study about the role of punishment in learning. I is instructed to ask S a number of questions; if the answer is incorrect, I is supposed to deliver an electric shock to S. The generator has 30 switches in 15-volt increments ranging from 15 up to 450 volts. I is supposed to increase the voltage each time S gives a wrong answer. In fact, there are no electric shocks; S is an actor who, although never actually harmed, shows increasing manifestations of pain as the voltage is increased.

The experiment was repeated in five different forms with respect to the closeness between I and S (Milgram 1974).

(1) In the "remote" setting, S is placed in another room and no vocal complaint is heard from him. However, at 300 volts, the laboratory walls resound as if pounded by S. After 315 volts the pounding ceases and no further answers are given by S (Fig. 8.1, situation 1).
(2) In the "voice feedback" setting, S is again in an adjacent room but his complaints can be heard by the teacher (Fig. 8.1, situation 2).
(3) In this setting S is in the same room as I which gives the possibility of visual contact (Fig. 8.1, situation 3).
(4) In this situation S and I sit side by side. At the 150-volt level, S refuses to place his hands on the shock plates (schematized by the black rectangles in Fig. 8.1, situation 4). The experimenter then orders I to force the subject's hand onto the plate. In this way, the experiment leads to a physical contact between I and S.
(5) This setting is similar to the previous one except that E is no longer in the same room and gives his instructions by telephone (Fig. 8.1, situation 5).

Figure 8.2 shows that the percentage of people who agree to carry the experiment to its termination (i.e. 450 volts) decreases when the "proximity" between

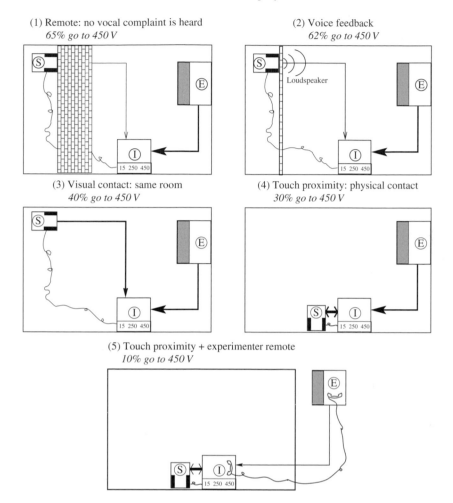

Fig. 8.1 Milgram's obedience experiment. *E*: Experimenter, i.e. the person who conducts the experiment; *I*: Instructor, i.e. the person who asks questions and delivers electrical shocks; *S*: Subject, i.e. the person who tries to provide answers and is supposed to receive electrical shocks.

In situation 1, the instructor and the subject are in two separate rooms without any communication between them except for an electrical wire through which the subject's answers are transmitted. In situation 2 they are also in separate rooms but the instructor is able to hear the reactions of the subject. In situation 3, the instructor and the subject are in the same room but several meters apart. In situation 4, they are close enough to touch one another. Situation 5 is identical to situation 4 except that the experimenter is located in a separate room. *Source: Milgram (1974).*

I and *S* increases. The experiments (especially 4 and 5) also suggest that there is a competition effect between the respective influences of *E* and *S* on *I*. This conflict is shown very clearly at some critical junctures in the course of an experiment by the exchanges taking place between the three persons. Consider

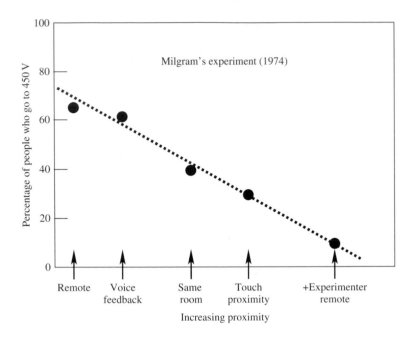

Fig. 8.2 Percentage of instructors who inflict 450 V shocks. The five data points correspond to the five experimental settings described in Fig. 8.1. For each situation the vertical scale gives the percentage (based on a sample of 40 different instructors) of those who pushed the experiment to its conclusion. *Source: Milgram (1974).*

for instance the following dialog which occurs after a 190-volt shock had been delivered (Milgram 1974).

> *S* (yelling): Let me out here. My heart is bothering me.
> *E*: The shocks may be painful but they are not dangerous. Continue please.
> *I*: You see he is hollering. No Sir, I'm not going to kill that man.
> *E*: The experiment requires that you go on.
> *S*: I absolutely refuse to answer anymore. Get me out.
> E: If the subject does not answer in a reasonable time, about 4 or 5 seconds, consider the answer wrong. Continue, please.

The dialog indicates that *I* is torn between the pleas of *S* who wants him to stop and the demands of *E* who encourages him to continue. The experiment provides a way of estimating the respective strength of these links. As often in the social sciences the main problem is to minimize the "noise". In this respect one must recall that the same instructor could not be used twice for after the experiment he (or she) was told about the real meaning of the experiment. As different instructors do not have the same reactions, the *ceteris paribus* condition can only be fulfilled on average for a sufficiently large set of instructors. Milgram indicates that 40 adults were studied in each of the settings schematized in Fig. 8.1.[2] It is because it is based on a large sample of experiments that the graphic in Fig. 8.2 is not completely obscured by variability and noise. If there had been only five experiments in each situation, the noise due to individual variations would have been too large to show any definite pattern.

There is one point which remains somewhat unclear, namely the exact significance of the horizontal scale in Fig. 8.2. In his book of 1974, Milgram labels this axis as showing "increasing proximity". But how should the term "proximity" be understood exactly? The simplest interpretation would be spatial proximity. It is true that between situations 1 and 4 the distance between I and S decreases, but this is clearly not the main factor. The real difference is the fact that the interaction becomes stronger because it is channeled through more and more means of communication: (i) almost no audio contact; (ii) audio contact; (iii) audio and visual contact; (iv) audio, visual and physical contact. However, in contrast to distance, these conditions cannot be expressed numerically. Thus, in a sense, the way the different situations are arranged on the x-axis relies to some extent on common sense knowledge. In physics the strength of an interaction can be expressed in joules. Figure 8.2 is certainly a big step in the right direction but it does not provide a completely objective picture.

In subsequent sections we give several historical illustrations of Milgram's law.

8.3 Effects of absentee ownership in Ireland

In the mid nineteenth century, during Queen Victoria's reign, Britain was the world's dominant power. British goods were shipped in British ships to all parts of the Empire as well as to the rest of the world. Yet Ireland was in a state of backwardness which it is difficult to fathom nowadays. Alexis de Tocqueville describes Ireland in the following terms (Tocqueville 1835).

The bed of a torrent seemed to be the only street in the village. I could not help remarking what I had seen so many times in Ireland. All the houses were of sun-baked mud made into walls to the height of a man; the roofs were made of thatch that was so old that the grass which covered them could not be distinguished from the grass of the neighboring hillsides. The houses had neither windows nor fireplaces. Light came in and smoke went out through the door. Inside a tiny peat fire burned slowly between four flat stones.

In 1875 the literacy rate (ratio of pupils to population of the 5–14 age group) was 32% as compared to 75% in France and Germany (Flora *et al.* 1983,

---

[2] This means that $5 \times 40 = 200$ experiments were carried out. If each experiment took about three hours (one hour for the experiment itself, one hour for the debriefing of I and one hour to record the results) this represents about 30 weeks of experiments. Moreover, the aspect considered in Fig. 8.1 was only one of the facets studied by Milgram's team. At that time Milgram was an assistant professor at Yale; it is remarkable that he got the funding to carry out such an ambitious project.

pp. 583, 592, 597). As we know, the period from 1835 to 1880 was marked by a series of famines which through malnutrition, disease, evictions and emigration reduced the population of southern Ireland (that is to say the land which now forms the Republic of Ireland) from 6.53 million in 1841 to 2.97 million in 1926.[3] The demographic catastrophe may have been the consequence of a combination of several factors, but the state of backwardness can be traced back to the absentee landlord system. This statement is based on the fact that it produced similar results in all countries in which it dominated, e.g. Tsarist Russia,[4] British India, Spanish Latin America and in a general way all colonies. Why does the absentee landlord system produce poverty and stagnation? The poverty is brought about by the capital drain and the stagnation arises from the segmentation between ownership and management. An illustration of the first effect is provided by the capital drain in Ireland. As we know, the big landowners of Irish estates spent most of their time in England. The resulting annual capital drain between 1700 and 1773 was estimated at around one million pounds (Lecky 1892). To put this figure in perspective one should recall that in 1735 the public income of Great Britain was 5.6 million pounds (Mitchell and Deane 1971, p. 387).

The second effect is more subtle. As will be seen in the next section, segmentation prevents the system from working properly and effectively.

8.4 Effect of segmentation on the effectiveness of a social system

By the expression "social system" we mean an organization which is set up to fulfill a given role. To begin with, we consider again the case of landowners and tenants. As the estates of absentee landlords are put in the hands of stewards it is not immediately obvious why this system should not work. After all, the steward has the knowledge, the authority and capacity to carry out sound management. Why should this kind of administration lead to stagnation? The following episode (Mingay 1956) gives an inkling of the kind of behavior which spoils the relationship between landowners and their stewards.

Sir Jacob Bouverie's estates in Kent totaled 1,200 hectares and brought in some 1,000 pounds a year. Although he was an absentee landlord who spent most of his time in London, Sir Jacob nonetheless exercised a close supervision over the way his steward managed the estate. The steward was constantly instructed to inspect

---

[3] Some historians asserted that with a population of 6.5 million, southern Ireland was obviously overpopulated. However, at 95 people per square kilometer, its population density was smaller than that in Northern Ireland (121 per sq kilometer) or England and Wales (106 per sq kilometer).

[4] By a twist of irony, several revolutionaries were themselves absentee landlords. A case in point is Alexandra Kollontai who held important positions after the Revolution of 1917. Although she never set foot on her domain, it was thanks to its revenue that she was able to travel throughout Europe, devoting all her energy to the destruction of the social order on which her own livelihood rested.

the farms, to keep down the rent arrears and on occasion to canvass those tenants who had the right to vote in favor of the parliamentary candidates supported by Sir Jacob's faction.[5] One episode gives an insight into Sir Jacob's damaging and inept interferences. The estate included the fishing town of Folkestone and in 1720, a February gale destroyed the harbor breakwater. Large rocks from the breakwater were washed onto the beach preventing fishermen from launching and beaching their boats. Sir Jacob's response was to offer the fishermen a tub of strong drink if they would put the stones back themselves. This offer, the fishermen evidently regarded as utterly inadequate. Henry Barton, the steward, replied that they were clamorous for more assistance and explained that the rocks could only be secured in position by large timbers. Sir Jacob received this opinion with indignation and felt that the tenants and the steward encroached on his sphere of authority as "the lord of the estate".

One might think that presenting one case is not sufficient to make the point. However, the bottom line is that for such a system to work well one must assume landowner and steward to behave in a way which is highly unlikely. The system could work if the landowner has total confidence in his steward and if the steward is totally honest and competent. However, in such a situation the steward would not subject to any control be: there is no control from below because the tenants have no say and there is no control from above because the landowner does not interfere. Observation shows that this is a very unstable equilibrium which can be broken by any exogenous factor, for instance an opportunity to earn extra revenue.

It takes only a small stretch of imagination to draw a parallel between the previous situation and the one corresponding to Milgram's experiment. More specifically the correspondence would be as follows: Landowner ⟷ Experimenter, Steward ⟷ Instructor, Fishermen ⟷ Subject. The experimenter's only objective was to carry out the experiment at all cost and he pursued it stubbornly. Similarly, Sir Jacob's only objective is to keep expenses down and to collect all the rent. Although in appearance he seems to have some control over the subject, the instructor is in fact pitifully powerless, squeezed as he is between the demands of the experimenter and the feelings inspired by the subject. Similarly we see that the steward is torn between the incessant but largely irrelevant demands of the landowner and the representations of the fishermen. The system does not work because the landowner has the power without the knowledge whereas the steward has some knowledge but little power; moreover, the tenants are poorly motivated because they have almost no say in their own

---

[5] By 1730 the number of voters in parliamentary elections represented about 1% of the population.

affairs and know that very little of the money they may earn will remain in their hands anyway.[6]

It could be argued that the landlord in our example was particularly inept and that the system may work better with more sensible landlords. After all the landlord $\longrightarrow$ steward $\longrightarrow$ tenants hierarchical structure may seem similar to the command structure in an army. Is it not possible to consider the landowner as a colonel, the steward as a lieutenant and the tenants as soldiers? The following observations show that there are fundamental differences between the two situations.

- The colonel has superior knowledge of the tactics that should be used against the enemy. He knows their strength, fire power, means of observation. Moreover, in most armies, he started his career as a lieutenant. In some armies officers are even required to serve as soldiers for a while. In short, the colonel has the capacity to give adequate and effective orders.

- In contrast, the absentee landlord does not know anything about farming or fishing. The only language that he understands is in terms of expenditures, income, debt, profit, etc. The same observation holds for investment funds, the modern analogue of the absentee landlords. As of June 30, 2005, State Street Corporation owned 11.2% of the shares of Boeing but State Street executives do not hold engineer degrees in aeronautics.[7] One could argue that ability to read Boeing's balance sheet is sufficient to implement the most cost effective options. That may be true as far as incremental innovations are concerned but not for major innovations. When Boeing started the 747 project in the 1970s it was a giant leap into the unknown. As we know, the project was an outstanding success, but the number of uncertainties (e.g. the price of oil, the rate of market development, the needs of passengers, etc.) was just too large to authorize definite predictions of success or failure.

- In the most effective armies the officers and even the generals lead their troops in battle. At the battle of Austerlitz (December 2, 1805) 14 French generals were killed or wounded (1 killed, 13 wounded). At the battle of Waterloo (June 18, 1815) 12 generals were killed or wounded (2 killed, 10 wounded) on the side of the Allies and 34 generals were killed or wounded (7 killed, 27 wounded) on the French side (Bodart 1908, pp. 369, 487). These figures clearly show that the parallel between generals and absentee landlords is not correct.

8.5 Hardship as a side effect of absentee landlordism

In the episode presented above, the landlord and steward do not harm the tenants directly. In Ireland, it was a completely different situation. It has been estimated

---

[6] For instance in Ireland the efforts of the tenants to increase the yield of their land were discouraged by the threat that any increase in yield and land value would bring about a rent increase.

[7] In fact 67% of the shares were in the hands of investment funds; State Street Corporation was the largest shareholder but other funds such as AXA, Capital Research and Management, and Barclays Global Investors were also important shareholders.

that over 250,000 people were forcibly evicted between 1849 and 1854. While the 1850s marked the climax of the crisis, the evictions continued in subsequent decades. Between 1879 and 1881 there were annually 11,000 evictions. In 1880 the Irish Catholics owned less than 5% of the land while 48% was held by the top 1% of landowners. The Gini coefficient[8] g of the concentration of land property was equal to 0.93, a level rarely seen elsewhere in the world (Guiffan 1989, Roehner and Rahilly 2002). The way the evictions were carried out is of interest from our perspective in this chapter. Under a law passed in 1847, called the "Gregory Clause", no tenant holding more than a quarter of an acre (0.1 hectare) of land was eligible for public assistance. Once tenants were formally evicted, the standard practice of the landlord's bailiffs was to level or burn the dwellings as soon as the tenants' effects had been removed. As a gesture of good will, the British Parliament passed a law which made it a misdemeanor to demolish a dwelling while the tenants were inside and prohibited evictions on Christmas Day and Good Friday (Campbell 1995, Donnelly 1995, Poirteir 1995). Usually, the evictions took place in the presence of a large number of Irish people and under the surveillance of a massive force of constabulary and military. The following excerpt from *The Times* (June 15, 1887, p. 12) suggests the bitterness of the rift between the two parties.

The eviction involved a constabulary force of about 100 men preceded by a guard of Royal Welch Fusiliers. Colonel Turner, the commanding officer, announced that he would deal very decidedly with any persons obstructing the police or throwing hot water on them.

Broken down by privation and exposure to the elements, the evicted people died by the roadside or tried to seek refuge in a workhouse. In an effort to "solve" the problem on his estates, Lord Palmerston resorted to forced emigration. In October 1847, his bailiffs put 177 of his tenants on a ship bound for Canada. The emigrants were so undernourished and poorly clothed that over a quarter of them died during the voyage.

From top to bottom, government, Parliament, landlords, stewards, bailiffs, constabulary and military seem to have acted in a particularly ruthless way. Nowadays, such abuses would probably qualify as crimes against humanity. But, just as with the instructors who inflicted 450-volt shocks in Milgram's experiment, this behavior can best be explained by a total lack of interaction between the Irish Catholics and the British. Although in England the antagonism between Catholics and Anglicans had lost a good deal of its earlier bitterness, in Ireland the gulf between the two communities was wider than ever. This is shown in a qualitative way by numerous

---

[8] The case when all individuals own the same share of land corresponds to $g = 0$, while $g = 1$ corresponds to the situation in which one landowner owns the totality of the land (with the remaining people being deprived of land). In 1880 the distribution of landownership in Ireland was remarkably close to this situation.

testimonies. In 1835, a lawyer in Dublin declared to Tocqueville: "Believe me when I say that I have dined only once in the house of a Catholic and that was by accident. Even for Catholics who become rich, Protestants cannot bear to see them on the same footing." It is remarkable that these lines were written almost a decade before the beginning of the great crisis. Other examples of an absence of interaction in similar situations can be found in Roehner and Rahilly (2002, pp. 227–9).

8.6 The absentee landlord paradigm in history

It is not possible within the limits of this chapter to give a systematic account of the role played by the absentee landlord paradigm. We will restrict ourselves to mentioning a few typical cases.

- Collectively, through its dignitaries and monasteries, the Church was a major absentee landlord. In France before the Revolution, the Church owned about 15% of the land, in Bavaria the monasteries were lords to 28% of all peasants. It is estimated that across Catholic Europe, monasteries owned about 10% of the land (Beales 2003, p. 3). Landownership by the Church was not specific to Catholic countries but was also common in Anglican countries, particularly England and Canada (see Wade 1832), in Islamic countries (see Keddie 1981, Clot 1990) or in Buddhist countries (for the case of Japan see Mason and Caiger 1973).
- Dukes, counts, earls and other members of the high aristocracy possessed huge estates. In 1780 the estates of the Duke of Orléans, who was the king's cousin, represented a total area of 24,000 square kilometers, which amounted to 5% of the territory of France (Lever 1996, p. 237). These estates were a mosaic of lands which were acquired, sold or transferred (e.g. through inheritance or dowry purposes) together with their tenants.
- The absentee landlord system was to be found in most colonies, including in some parts of the Thirteen American Colonies. For instance, the colony of Maryland was given to George Calvert, Lord Baltimore, by King Charles I in 1632. In addition to Maryland the colony also covered at that time parts of Delaware and Virginia. Frederick Calvert (1731–71), the sixth Lord Baltimore, was proprietor of Maryland from 1751 until his death at Naples in Italy but never set foot on the soil of his American estate. In 1663, King Charles II gave a large tract of land including present-day North Carolina to some of his friends: General George Monck, Sir George Carteret, who was one of the wealthiest men in England, and Sir William Berkeley, who had been a ruthless governor of Virginia. Only William Berkeley ever set foot on Carolina soil.
- In Russia since Peter the Great all male members of the Russian nobility had to serve in the military or civil service without regard for individual preference; moreover, whatever their rank in the nobility they could not immediately obtain a high level position. During his six-month reign (1762), Peter III abolished this service obligation; as a result, the aristocracy became even more estranged from the rest of the society.
- The notion of absentee ownership can be extended to the case where foreign landlords control the financial resources of a region or of a whole country. For instance, in the

second half of the nineteenth century the Chinese Maritime Customs revenues were collected by the British. The fact that many railroads were owned by foreign companies also contributed to the control of the economy by the different foreign powers who were granted economic privileges and concessions. This form of economic colonialism eventually led to a long period of civil wars in which warlords backed by competing foreign interests opposed one another (Goetzmann and Ukhov 2001, Treat 1928).

The fact that a landowner does not visit his estates cannot be regarded as an absolute proof of a lack of interest. In the next section we define another criterion which applies particularly to settler colonies.

8.7 Assessing interaction in settler colonies

In the previous sections we argued that a lack of interaction between rulers and their subjects has two concomitant effects: (i) it makes the ruler insensitive to the suffering of the subjects; (ii) it prevents the ruler developing an understanding of the problems faced by his subjects. Previous historical examples suggest that, apart from the hardships suffered by the subjects, the most obvious consequence of this lack of interaction is technical stagnation and economic decline. So far, however, we have not offered an objective criterion for measuring the strength of interaction. This is difficult at the level of individual landlords but can be done for a large population of landlords, which is why, in this section, we are mainly interested in settlement colonies. Such situations involve two different populations, which makes it possible to use the criterion of intermarriage rates. A low intermarriage rate between settlers and the rest of the population points to a weak interaction and therefore signals a situation in which the absentee landlord syndrome may play a role. In contrast, high intermarriage rates suggest a situation in which the settlers have been able to blend into the population. It is well known that in Ireland, due to the Penal Laws, marriages between Roman Catholics and Anglicans were almost impossible. In what follows we illustrate the application of the intermarriage criterion by examining two very different cases: the colonization of the French province of Normandy by the Vikings and the colonization of Mexico by the Spanish.[9]

8.7.1 Colonization of Normandy by the Vikings

From the eighth to the eleventh century, the Danes were known as Vikings. Together with Norwegians and Swedes, they colonized, raided and traded in all parts

---

[9] In so doing we follow once again the methodology of the extreme value technique introduced in an earlier chapter.

of Europe. The Vikings temporarily conquered parts of England, known as the Danelaw, and France, giving the name to the French region of Normandy (Normandy comes from the French word *Normands* which designates the Vikings). Their raids in the Seine valley led them to Rouen and Paris. For instance, in 841 the city of Rouen was burnt down and important monasteries were looted, ransacked or held to ransom. In 845, Ragnar Lodbrok besieged Paris with 120 ships and 5,000 warriors. The king, Charles the Bald, agreed to pay them 7,000 pounds of silver in order to spare the city. Looting, burning and extorting ransoms is the standard behavior of invaders. Hernán Cortés's conduct in Mexico was no different.[10] After 880 the Viking presence in the Seine valley had become permanent. The monks had to flee the region seeking refuge deep in the countryside. Other Viking groups had similarly settled in England and Ireland. In parallel with what happened in Latin America, the outcome could have been a Viking empire with its center in Copenhagen. In many respects the parallel makes sense. For instance, the numbers of the Viking conquerors were of the same order of magnitude with respect to the population as in Mexico or Peru and they had an obvious military supremacy. Yet this did not occur. All of a sudden, something rather unexpected happened. In 911, Rollo, the leader of the Viking colony, started negotiations with the king, Charles the Simple, in order to formalize his sovereignty which already existed de facto. This move resulted in the treaty of Saint-Clair-sur-Epte in which the king gave up to the Vikings a territory extending from Rouen to the sea. In return Rollo accepted Christianity and agreed to marry the king's daughter Gisla. In 918 he married her in a Christian ceremony. Not only was his son William baptized and raised by clerics but even the children he had from a previous marriage with a Viking wife were also baptized. In short, the Vikings blended with the natives and adopted their religion. Their customs and language became mixed with local usage. The Scandinavian linguistic influence is still to be found in numerous Norman place names with endings such as —tot (farm), —thuit (cleared area), —bee (stream), —hogue or —hague (hill) or in family names such as Burnouf, Thouroude, Yngouf.

8.7.2 Colonization of Mexico by the Spanish

The conquest of Mexico by Hernán Cortés started very much in the same way as the conquest of Normandy by the Vikings. First, as we already mentioned, there was a phase of plunder. In a second phase Cortés took a native wife and learned the Aztec language. Several of his companions followed his example and

---

[10] Even more recently, the behavior of the Allied forces which invaded Peking during the Boxer War in 1900 was very similar.

intermarriage remained the rule until 1529. At this time, Cortés came back to Spain where, mainly for political reasons, he had to take a wife in the Spanish nobility. This marriage allowed him to keep his estates in Mexico but in 1542 the ownership of the domains of the conquistadors was transferred to the Crown. Many other things changed at the same time. The Franciscans who had been favored by Cortés and had developed fairly close ties with native people were replaced by the Dominicans who had a much more rigid approach; it should be remembered that the Dominican Order was closely associated with the establishment of the papal inquisition. Not only did Cortés never adopt the religion of the Aztec but on the contrary the Spaniards demanded that the idols be removed and that shrines of the Virgin Mary be set up in their place. In short, 23 years after the conquest of Tenochtitlan, the Aztec capital of Mexico, the country became dominated by absentee landlords.

The Spanish conquest of other Latin American colonies followed the same pattern. As a matter of fact, when Nikolaus Federmann (in Venezuela), Francisco Pizarro (in Peru and Bolivia) and Hernán Cortés came into contact with the peoples of Latin America, they found vibrant civilizations characterized by impressive architectural achievements and whose craftsmanship won the admiration of Spanish courtiers when they discovered the magnificent items sent back by the conquistadors. Unfortunately, after the conquest the Inca and Aztec civilizations withered, regressed and eventually collapsed. The role played in this process by absentee ownership becomes more evident when one realizes that the main objective of Federmann, Pizarro or Cortés was not military conquest and settlement but, as stated in a letter from Charles V to Cortés dated June 26, 1523, "to extract rents from the new territories" (Duverger, 2001). Actually, the territory of New Granada, which corresponds approximately to present-day Venezuela, was granted by Charles V to the Welsers, the great Augsburg banking firm to which he was heavily indebted. Federmann was an agent of the Welsers and their government was marked by ruthless exploitation of the Indians (Langer 1968, p. 529). These objectives were very different from the goals pursued by the Vikings in Normandy.

How can one explain why the behavior of the Spaniards was so different from the attitude of the Vikings? Perhaps the main reason is that the Vikings were not backed by a powerful state as were the Spaniards. In this respect it is striking to see that even 5,000 kilometers away from their country, the conquistadors remained in contact with the king and the court. They appealed to the king to settle their disputes about boundaries, the king granted charters which set the respective rights of the conquistadors and of the king (both Pizarro and Cortés returned to Spain to seek more favorable charters). The conquest of the Philippines followed a similar pattern. This explains the remark made by General MacArthur about estates in the sugar region being owned by landlords in Madrid.

The explanation that we put forward may seem plausible, but the examination of other cases calls it into question. Nobody would deny that the Roman Republic and Empire were strong states. Yet the policy of the Romans after their conquest of Gaul was very different from Spanish policy in Latin America. There is a famous speech made by Emperor Claudius in AD 48 in which he asks whether or not people from the provinces (i.e. from outside Italy) should be admitted to the Senate. It is reported by Tacitus (Book 11, Chapters 23–24) in the following way.

You say: isn't an Italian senator preferable to a provincial one? If you consider all our wars, none lasted a shorter time than the one against the Gauls, and now that they have been assimilated by our customs, our culture and by intermarriage with us, let them bring their gold and wealth here. Look at that most splendid and prosperous colony of Vienne [a city in Gaul, 30 kilometers south of Lyons] and for how long it has supplied senators in this senate house. From this colony comes Lucius Vestinus, that adornment of the equestrian order [the equestrian order was somewhat similar to the English gentry, that is to say a nobility which was based on merit as well as on birth].

Not only was intermarriage well accepted, but the Gauls were admitted in the highest ranks of the Roman nobility. To the contrary, in Latin America and in Ireland it was segregation which dominated. This is well illustrated by the Statute of Kilkenny (1367); the following excerpt is very explicit in this respect.

Now many English of the said land [Ireland] forsaking the English language, manners, mode of riding, laws and usages, live and govern themselves according to the manners, fashion and language of the Irish enemies. It is ordained and established that no alliance by marriage, gossipred [i.e. sponsoring a child at baptism], fostering of children [i.e. being brought up in the household of another family] or concubinage nor in any other manner, be henceforth made between the English and Irish. Is is agreed and established that no Englishman be governed by Brehon law. If any do the contrary, he shall be taken and imprisoned and adjudged as a traitor.

These sentences are really surprising. If the Statute is not a forgery (a possibility that should not be discarded too quickly), it is really difficult to understand what motivated such strong language. It can be noted that the first Penal Laws against the Catholics were re-enacted in 1559 (Act of Supremacy and Uniformity), that is to say more than two hundred years later. In a sense the Penal Laws continued the tradition set by the Kilkenny Statute with the difference that the Statute was based on ethnicity whereas the Penal Laws were based on religion. Of course, we do not know to what extent the Kilkenny Statute was enforced nor do we know whether the picture drawn by Claudius really reflected the reality. However, these records seem to show two very different and almost opposite conceptions.

8.8 Revolutions seen as a way to end absentee landlordism

We already noted in a previous chapter that in a general way successful revolutions tend to increase social interactions, in particular by removing major obstacles to the establishment of bonds between various categories of citizens. It is not surprising therefore that one of the main purposes of revolutions is to get rid of absentee landlords. This is illustrated by the following examples.

- In many countries the Reformation brought about major changes in the distribution of power and wealth, in particular by the confiscation of ecclesiastical property. A list of cases for 10 European countries can be found in Roehner and Syme (2002, p. 119).
- A major outcome of the American Revolution was the confiscation of the property of Loyalists. It is true that not all Loyalists were absentee landlords, but probably all absentee landlords were Loyalists. As we have seen earlier, in 1772 members of the British aristocracy were in possession of vast estates.
- Two of the first moves of the French Revolution were to abolish rights of feudal lords and to confiscate the landed property of the Church.
- The insurrection which eventually led to the independence of Ireland marked the end of the power of the absentee landlords. It is true that there had been a land reform in 1903 but it resulted in a long-term debt that tenants had to repay over several decades. After independence, the remaining debt was unilaterally canceled by the Irish government.
- The expropriation of Japanese absentee landlords after World War II was a landmark reform which probably would not have been possible without the support of the occupation forces; this is why it can be seen as a revolutionary move.

In all these cases, ending the privileges of absentee landlords brought about more than social justice; it also improved economic efficiency and social cohesion.

8.9 Present-day manifestations of the absentee ownership syndrome

Absentee landlord situations are fairly common nowadays in developing countries (see for instance the case of Guatemala described by Dosal 1995), but are there situations of absentee ownership in the industrialized countries of the twenty-first century? We already mentioned the role played by investment funds as major shareholders in big corporations such as Boeing. The purchase of major industrial companies by investment funds brings us even closer to the standard absentee landlord situation. An example chosen almost randomly in the *Wall Street Journal* (November 29, 2005, p. 2) is described in the following announcement.

The Danish telecommunication operator TDC is close to an agreement to be bought by a consortium of five private-equity funds [of which two are British and three are American] for roughly $12 billion. The chairman of the TDC board declared that the bid was found very attractive by the shareholders.

Unions, to the contrary, expressed their worries about the company's future. Many feared a scenario in which the consortium would break up TDC, set up separate subsidiaries and sell them (at a profit) to various telecommunication competitors.

As explained earlier, when companies are controlled by absentee owners incremental innovations are preferred to bolder and more risky ones. In some industries this can lead to satisfactory financial returns, but in the long run it seems to be a good recipe for technological stagnation. At present this is no more than a prediction. One will have to wait for further evidence to see if the lack of interaction will produce the same results as those described in this chapter.

Different forms of the absentee landlord paradigm are summarized in Fig. 8.3. Establishing a connection between cases which at first sight seem to have little in common is a first step in proposing a comprehensive theory.

If one combines the mechanism described in the present chapter with the trend delineated in Chapter 6 one gets a picture of more and more segmented countries and societies. In the age of globalization this may at first seem a paradoxical statement. Yet there are strong trends which point this way. (i) The increasing remoteness, detachment and disconnection observed between ownership and

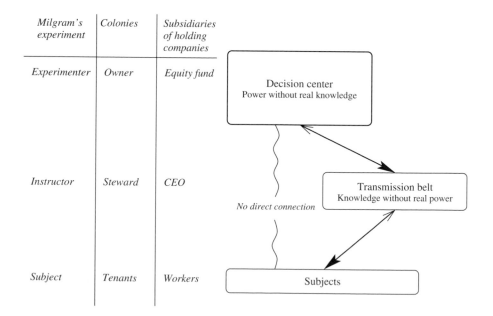

Fig. 8.3 Milgram's experiment and two different embodiments of the absentee landlord paradigm. Milgram's experiment demonstrates that the remoteness between the decision center (experimenter and instructor) and the subject makes the former more indifferent to the suffering of the latter. In addition, the absence of connection makes any bottom-up feedback impossible. Situations of this kind are described in Nearing and Freeman (1926) and Staley (1935).

employees, as documented in the present chapter. (ii) The fact that corporations are becoming less and less subject to oversight and accountability by states. (iii) The fact that, thanks to the growing role granted to tax havens,[11] the contribution of corporations to state income is dwindling. The normal outcome of these tendencies may be a kind of new Middle Age. It should be noted that the segmented structure of Europe in the Middle Ages in fact favored globalization, in particular through the dynastic connections which existed between members of the European aristocracy.

---

[11] During the period 1980–2002 the volume of bank deposits held offshore rose from virtually nothing to $11,000 billion (*Observer*, November 17, 2002). By 2002 almost all major international transactions involved capital movements through tax havens for purposes ranging from confidentiality and secrecy to tax evasion.

Part III

Micro-interactions: a network explanation of suicide

The statistical investigation of the phenomenon of suicide began in the mid nineteenth century when data about suicide rates became available in all industrialized countries. Emile Durkheim's book about suicide which appeared in 1897 was a landmark in this research but there had been many earlier studies.[1] The contribution of Durkheim stands out because he proposed a social interpretation of suicide rather than one based on individual, psychological factors. He maintained that one of the main factors of suicide is social isolation. According to this view it is the paucity of links and interactions with other people which leads to suicide. This perspective is similar to the interpretation of apoptosis according to which cell death occurs when the cells no longer receive the "stay alive" message from their neighbors (see the chapter on apoptosis).

Yet, to this day, the phenomenon of suicide remains fairly mysterious. Common sense seems to be misleading rather than helpful as illustrated by the two following observations. (i) Contrary to the assumption or expectation of many people, the monthly maximum of suicide rates does not occur in fall or in wintertime but in May or June. (ii) Even events as dramatic as the attack of September 11, 2001 have no visible impact on the number of suicides. This observation, which will be explained in more detail in a subsequent chapter, holds at monthly as well as at daily level. In other words, (i) the number of suicides in the United States in September 2001 is no different from the numbers observed in September 2000 or September 2002, and (ii) the daily numbers of suicides on September 11, 12 or 13 are very similar to the other daily numbers of September 2001.

Many empirical regularities have been suggested and studied but in fact very few of them have a broad validity in time and space. Let us give some examples of rules which were at first thought to hold until they were invalidated by exceptions and anomalies.

- **Impact of economic recessions** It is often claimed that suicide rates rise in times of economic recession, but there are in fact many exceptions. Overall the correlation between suicide and unemployment is fairly low. For instance, in Germany over the time interval 1881–1989, it is equal to 0.19. If one makes a distinction between males and females, the picture becomes even more confusing: it appears that over 1949–89, the suicide rate of males is positively (but weakly) related to unemployment, $r = 0.26$, whereas the correlation is negative (and even weaker) for females, $r = -0.16$ (Weyerer and Wiedenmann 1995).

---

[1] One can mention for instance the following studies (listed chronologically): Boismont (1865), Le Roy (1870), Cristau (1874), Morselli (1879), Legoyt (1881), Masaryck (1881), Nagle (1882). More recent studies include (in alphabetical order): Bayet (1922), Botte (1911), Candiotti *et al.* (1948), Chesnais (1976), Hezel (1987, 2001), Lester (1997), Olson (2003), Ropp *et al.* (2001), Rubinstein (2002), Schimdtke *et al.* (1999), Zacharachis *et al.* (2003).

- **Male versus female suicide rates** In Western countries the suicide rate of males is about twice the suicide rate of females. Yet in China and some provinces of India, female suicide rates are higher than male rates.

- **Trend in male versus female suicide rates** Usually male and female suicide rates move in the same direction which means that they have a high positive correlation. Yet there are many exceptions to this rule. For instance, in the period 1974–81 Japanese male and female suicide rates were negatively correlated. This observation is particularly disconcerting because the observation of a strong positive correlation over many decades may be seen as reflecting a stable structural property and it is therefore difficult to understand what brings about such sudden switches.

- **Influence of population density** In the time of Durkheim, suicide rates were higher in big cities than in the countryside. Nowadays it is the opposite: suicide rates in Montana or Wyoming are about twice as high as in Boston.

- **Suicide rates in time of war** It has been argued that suicide rates decrease in time of war. While this seems indeed to hold for some countries and for some wars, there are many exceptions. US suicide rates declined during World War I and World War II, but they increased during the Korean and Vietnam wars; moreover, for World War II the decrease began as early as 1939. While both male and female suicide rates declined in Denmark during World War I they increased during World War II; thus, the female suicide rate more than doubled from 10 to 23 per 100,000.

- **Suicide rates in the armed forces** In the time of Durkheim, suicide rates in the armed forces were two or three times higher than in the general population. Nowadays, however, suicide rates in the armed forces, for instance in the US Army, are at the same level as in the general population.

All these exceptions and broken regularities may seem fairly disconcerting. However, one can observe that all the cases that we mentioned refer to anthropomorphic categories. "Being unemployed", "belonging to the armed forces", "living in time of war", all these notions have no clear and univocal interpretation in terms of network science concepts.[2] In contrast, the regularities that we examine in subsequent chapters are fairly robust and have a clear interpretation in network terms. As we will see there is a clear relationship between various forms of social isolation and suicide.

In the next two chapters we investigate the implications of the marital bond for suicide. Then we explore other situations of social isolation such as the conditions of inmates or immigrants. Finally, we discuss the phenomenon known in biology as apoptosis (often called cell suicide) in the light of the framework built up in the preceding chapters.

---

[2] To give them a well-defined meaning one would have to describe the notions of unemployment, armed forces or war in terms of networks, links and interactions. By doing so, one would discover that in network terms there are *different variants* of unemployment. For instance, a jobless person in a country with a high level of social solidarity such as the Scandinavian countries will keep more social links than would be the case in a country where unemployed people do not benefit from any social protection.

9

Effects of a male–female imbalance

In this chapter we examine the effect on suicide rates of being deprived of family ties with spouse and children. It has been known at least since the work of Emile Durkheim that unmarried people have a higher suicide rate than married people. It is our objective in the present and subsequent chapters to identify and test the various implications of this effect.

In normal conditions, it has little impact on the total suicide rate because in most societies unmarried people are a small minority; consequently, their higher suicide rate has only a small impact on total suicide rates. To be more specific, let us consider a population A of 100,000 people for which the suicide rate is 15 per 10^5 and which comprises only married people (for the sake of simplicity we suppose that there are no children). In such a population there are $n_A = 15$ suicides annually. How is this number modified in a population B similar to A in all respects except that it comprises 5% unmarried people? If we assume that the suicide rate of unmarried people is $15 \times 2 = 30$, the annual number of suicides will be $n_B = (15/10^5) \times 0.95 \times 10^5 + (30/10^5) \times 0.05 \times 10^5 = 15.75$, which represents a relative increase of $(n_B - n_A)/n_A = 5\%$. As might be expected (and will be seen in more detail subsequently) a change of this magnitude is likely to be hidden by the background noise of suicide rates. If we consider a third population C which is identical to A and B except for the fact that it comprises 50% unmarried people, we get $n_C = (15/10^5) \times 0.5 \times 10^5 + (30/10^5) \times 0.5 \times 10^5 = 22.50$ and $(n_C - n_A)/n_A = 50\%$. This calculation shows that a group of unmarried people will have an observable effect on suicide rates only if it represents about 50% of the population. This leads us to search for populations in which there is a high proportion of unmarried people. One can think of two mechanisms. (i) In populations characterized by strong gender imbalances some men (or women) will not be able to get married. This effect will be examined in the present chapter. (ii) For some reason, young people may postpone their marriage, which will increase the suicide rate in young age groups. This effect will be examined in the next chapter.

After reviewing the statistical evidence about increased suicide rates among unmarried people, we build a simple model based on this evidence which provides analytical formulas expressing suicide rates as a function of the males to females ratio r. In the rest of the chapter we set up quasi-experiments to confront these predictions with actual observations. Finally, once we have been able to convince ourselves that these formulas provide a fairly correct description, we discuss some of their possible implications, for instance the effect on suicides of the death of young males during major wars.

9.1 Suicide rates of unmarried versus married people

If, as implied by the network representation, the phenomenon of suicide reflects the sparseness and weakness of the ties which link up individuals with their society, then the lack of the marital bond should have a major and fairly universal effect on suicide rates. Table 9.1 summarizes the evidence for four different countries and for time intervals which cover more than one century. On average, the suicide rate for unmarried men is 2.3 times higher than for married men; for women the ratio is 1.9. Subsequently, these ratios will be denoted by k_m and k_f respectively.

An objection may be raised which we must discuss. Suicide rates are known to be age dependent with higher rates in older age. As, in addition, the mean ages of single and married people are not the same, one may wonder if the ratios measured in Table 9.1 are not statistical artifacts. Figure 9.1 answers this question. It shows that in every age group the suicide rates of unmarried people are higher than those of married people. In addition, we see that the ratio is largest for the age interval corresponding to the family-building period of life, namely 30–45 years. During this period, married people also have close ties with their children.

One century ago, a substantial proportion of the population was living and working on farms. It was a way of life which implied fairly different roles for men and women. For instance, unmarried women were likely to stay on the farm with their parents, at least if the farm was big enough to provide a living for all. In contrast, young men were supposed to get married and to set up their own farms. However, in the course of time there was a gradual population shift to the secondary and tertiary sectors coupled with an increasing female employment rate. Under such conditions one would expect the lives and social networks of unmarried females and males to become more similar. This hypothesis can be tested statistically. If one computes the suicide rates of unmarried males and females one observes that there is a convergence process in which the rate for unmarried males strongly decreases while the rate for unmarried females

Table 9.1 *Influence of family ties on suicide rates*

| | Time interval | Country | Ratio of suicide rates: unmarried/married | |
			Males	Females
1	1889–91	France	2.80 ± 0.23	1.56 ± 0.43
2	1881–90	Switzerland	1.66 ± 0.24	1.34 ± 0.41
3	1911–20	Norway	2.56 ± 0.83	2.18 ± 1.21
4	1968–78	France	2.69 ± 0.66	2.24 ± 1.10
5	1970–85	Norway		1.78 ± 0.24
6	1981–93	France	2.34 ± 0.33	2.15 ± 0.91
7	1982–96	Britain	1.44 ± 0.12	1.27 ± 0.20
8	1990–92	Queensland	2.67 ± 1.20	2.11 ± 1.60
9	1998	Australia	2.21	2.00
Average			2.30 ± 0.24	1.85 ± 0.32

Notes: In most cases (except 9) detailed data by age interval were available. We computed the ratios of the suicide rates in each age interval and then the average m and the standard deviation σ of the ratios. The results in the table are given in the form $m \pm \sigma$. It should be noted that the category "married" includes both "married without children" and "married with children"; consequently, the reduced suicide rates for "married" should not be attributed solely to marriage but to the combined effect of being married and having children. It can be noted that the figures for Britain and Switzerland are curiously out of range.

Sources: 1: Durkheim (1897); 2: Halbwachs (1930); 3: Statistiske Centralbyrå (1926, Table 22); 4: Besnard (1997); 5: Hyer and Lund (1993); 6: Besnard (1997); 7: Kelly and Bunting (1998, Fig. 4); 8: Cantor and Slater (1995); 9: Steenkamp and Harrison (2000).

Table 9.2 *Suicide rates of unmarried people: male vs. female*

	1889–91	1968–73	1979–83	1989–93
Unmarried males, 30–59 y	101	65.4	76.5	65.8
Unmarried females, 30–59 y	16.7	17.2	24.0	23.1
Ratio M/F	6.0	3.8	3.2	2.8

Notes: The table shows that over the past century suicide rates of unmarried males and females have become closer. There was a similar, albeit slower, trend for the rates of married men versus married women: for the 30–59 age interval the ratio decreased from 3.20 in 1889–91, to 2.45 in 1989–91.

Source: Besnard (1997).

increases slightly. The convergence documented in Table 9.2 suggests that over the past century the social networks of males and females have become more similar. Because these people are deprived of marital ties, which are normally the predominant factor, these rates enable us to probe the second circle of social ties.

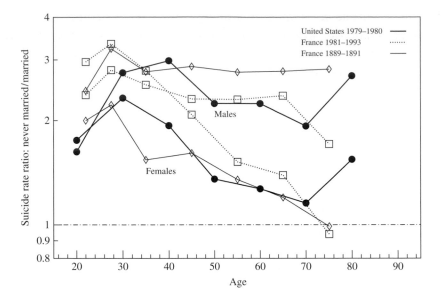

Fig. 9.1 Ratio of suicide rates: unmarried/married. All ratios in all age groups (under 75) are higher than 1 which means that the suicide rates of unmarried people are higher than the suicide rates of married people not only globally, but also in each age group. Moreover, in a general way, the ratio is higher for males than for females; the only exceptions are the first three points of the curve for France 1981–93. For both men and women the ratio is largest during the age interval 30–45. *Sources: United States: Vital Statistics of the United States, 1979 and 1980, Vol. 2, Part I, Section 1, p. 323; population of age groups: Statistical Abstract of the United States (1981, p. 38); France 1981–93: Besnard (1997); France 1889–91: Durkheim (1897, Table 21).*

This convergence is less visible for married people because in this case it is the permanence of domestic ties which is the dominant factor.

In a more general way, we can retain the idea that if one wishes to observe the effect of changes which occur outside of the family unit, it is a good strategy to study how they affect the suicide rates of unmarried people (for instance in the age group 15–24) rather than the rates in the general population in which married people are predominant.

9.2 Suicide rate in a population with a gender imbalance

We consider a population in which the male and female populations, which we denote by M and F, are not equal: $M/F = r \neq 1$. What is the effect of this imbalance on the suicide rate of males (t_m), females (t_f) and on the total suicide rate (t)? We denote by $\bar{t}_m, \bar{t}_f, \bar{t}$ the same rates in the case when $r = 1$. From the previous section we know that there will be an increased suicide rate for the men and women who remain single. Throughout this chapter we assume that the male–female imbalance is the only limiting factor on the number of marriages; such an assumption is acceptable when r is substantially different from 1, say $r > 1.5$, because in this case the number of unmarried people due to the imbalance is notably larger than the number of people who do not get married for other reasons.[1]

Using a simple argument which is detailed in the Appendix, we find

$$t_m = \begin{cases} \bar{t}_m & \text{if } r \leq 1 \\ (1/r)\bar{t}_m + (1-(1/r))\, k_m \bar{t}_m & \text{if } r \geq 1 \end{cases} \tag{9.1}$$

Let us see how this expression can be interpreted. When $r \leq 1$ all men are married and it is therefore normal that their suicide rate is given by $\bar{t}_m$. When $r \geq 1$, t_m comprises two terms: (i) the term $(1/r)\bar{t}_m$ describes the suicides of the M men who are married; (ii) the term $(1 - 1/r)k_m\bar{t}_m$ describes the suicides of the $M - F$ men who are unmarried; it is natural that for these men we have to use the suicide rate $k_m\bar{t}_m$ which characterizes unmarried people.

There are similar expressions for t_f and t (see Appendix):

$$t_f = \begin{cases} r\bar{t}_f + (1-r)\, k_f \bar{t}_f & \text{if } r \leq 1 \\ \bar{t}_f & \text{if } r \geq 1 \end{cases} \tag{9.2}$$

$$t = \begin{cases} [2r/(r+1)]\bar{t} + [(1-r)/(1+r)]\, k_f \bar{t}_f & \text{if } r \leq 1 \\ [2/(r+1)]\bar{t} + [(r-1)/(r+1)]\, k_m \bar{t}_m & \text{if } r \geq 1 \end{cases} \tag{9.3}$$

How many parameters do we have in this model? It can be noted that $\bar{t} = (\bar{t}_m + \bar{t}_f)/2$, as a result there are five parameters: $r, k_m, \bar{t}_m, k_f, \bar{t}_f$. Naturally, these parameters are not free parameters. In any population for which the male and female populations are known, r is determined. Furthermore, k_m and k_f must take values which are consistent with those in Table 9.1. Finally, if the same population can be observed not only for $r \neq 1$ but also for r close to 1, this observation will give estimates for $\bar{t}_m$ and $\bar{t}_f$. In such a case there are no free parameters in the model.

Formulas (9.1)–(9.3) are illustrated in the graphs of Fig. 9.2. Let us for instance consider the section $r \leq 1$ of the middle curve of the total suicide rate; why is it a horizontal line? As r becomes smaller, there are more and more women in the population and this has two opposite effects: (i) as men are replaced by women, the suicide rate is reduced because women have a smaller rate than men; (ii) more and more women remain unmarried and these women have an inflated suicide rate with respect to married women. When these effects are of same magnitude they

---

[1] In this chapter we consider mainly data for periods prior to 1960, a time when divorce rates were less than 3%. This is why the incidence of divorce can be neglected. It will be taken into account in the next chapter.

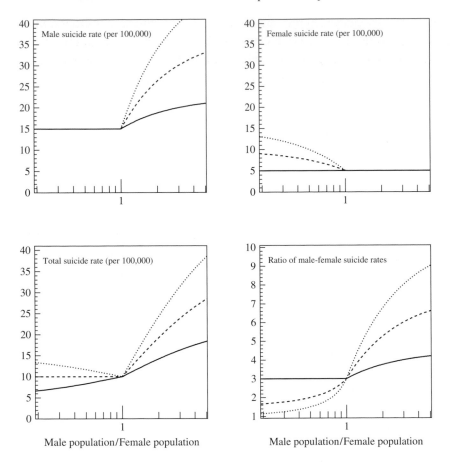

Fig. 9.2 Suicide rate as a function of sex ratio. The graphs show the suicide rates given by formulas (9.1)–(9.3). All curves correspond to $\bar{\imath}_m = 15, \bar{\imath}_f = 5$, but to different values of the coefficients k_m and k_f. For the lowest curves $k_m = 1.5, k_f = 1.0$; for the middle curve $k_m = 2.5, k_f = 2.0$; for the upper curves $k_m = 3.5, k_f = 3.0$.

cancel each other with the result that the curve is horizontal. Such a cancellation occurs when $\bar{\imath}_m/\bar{\imath}_f = 2k_f - 1$.

In the next section we examine how the model described by Formulas (9.1)–(9.3) can be tested.

9.3 Designing the experiment

In a previous chapter we emphasized that one of the main difficulties in the social sciences is the high level of "noise". By this term we understand the influence of all factors apart from the effect that one plans to observe. More precisely, if the signal that one tries to identify has an amplitude which is of the same magnitude as (or smaller than) the standard deviation of the noise, one is in a situation in which it is difficult to draw any clear conclusion. Thus, the main requirement is to find situations for which the signal is the strongest possible. In order to illustrate this important (but often overlooked) point we first describe two experiments which were *not* successful. Nevertheless, they are probably not without interest because they show that it is only by a process of trial $\longrightarrow$ failure $\longrightarrow$ improvement that it is possible to progressively enhance the signal to noise ratio.

9.3.1 An unsuccessful experiment

We will try to test formula (9A.3) for the ratio t_m/t_f. This ratio has the advantage of depending on only three parameters (apart from r). For this test we need a database which gives (i) suicide rates by gender, and (ii) male and female population for determining the ratio $r = M/F$. WONDER is a database created and maintained by the Centers for Disease Control of the American Department of Health. This database in fact gives mortality rates for many causes of death; in the ICD-9 classification which was in use in the 1980s the code for death by suicide was 950–959. WONDER gives suicide data nationally, and also at state and county level. It would seem therefore that we are in a good position to carry out this investigation. If one remembers that there are 3,000 counties in the United States it is clear that we have a huge amount of data at our disposal. Moreover, the fact that the database covers more than 20 years gives us the possibility to carry out longitudinal (i.e. in the course of time) as well as cross-sectional (i.e. across states or counties) analysis. Because r does not change markedly over the 20 years covered by the database it makes sense to try a cross-sectional analysis; because the number of suicides in many small counties would be too small, we make the cross-sectional analysis at state level. In the hope of reducing the dispersion, we ask WONDER to perform averages over the five years 1979–83. The main advantage of summing up the suicides over five years is to increase the number of suicides in states which have a small population (New Hampshire, Rhode Island, etc.) and therefore to reduce statistical fluctuations.

Figure 9.3 shows the results of the investigation for the 50 states and the District of Columbia. The points in the scatter plot are contained within a narrow interval $r \in (0.85, 1.15)$ but they have a great vertical dispersion in rate ratios. The coefficient of correlation turns out to equal 0.20, which is a low and in fact non-significant correlation: the confidence interval for probability 0.95 is $(-0.08, 0.45)$. The thin line shows the theoretical curve given by the expression (9A.4); the fact that it crosses the center of the scatter plot has been obtained by choosing the parameter $\gamma = \bar{t}_m/\bar{t}_f$ equal to 4 and k_m, k_f equal to 3.4 and 1.8 respectively. Unfortunately, the relationship that one wishes to observe is drowned

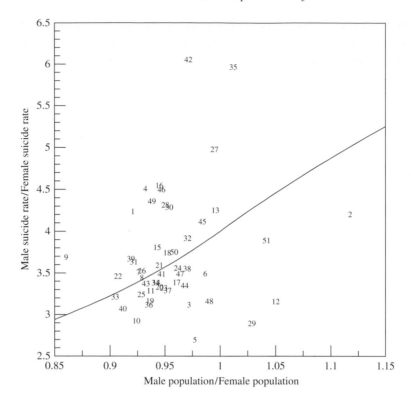

Fig. 9.3 A failed attempt to test Formula (9A.4). Formula (9A.4) gives the ratio of the male to female suicide rates as a function of the sex ratio r for 50 American states and the District of Columbia. The three parameters of (9A.4) have been chosen in such a way that the thin line crosses the center of the scatter plot, namely $\bar{\tau}_m/\bar{\tau}_f = 4.0$, $k_m = 3.4$, $k_f = 1.9$. However, the dispersion of the data points is too large to allow any clear conclusion. *Sources: WONDER database (website of the Centers for Disease Control).*

in the background noise. Obviously, our first attempt failed miserably; it is of interest to understand why.

The main hurdle in the previous attempt is the fact that for 90% of the states r is between 0.90 and 1.00. Remember in this respect that the effect is weaker for $r < 1$ than for $r > 1$; there are only five states for which r is larger than 1, and even for these r is very close to 1. As a result the effect that we wish to observe is very weak. In addition, what makes the scatter plot of little usefulness is the huge dispersion. What are the dispersion factors? The ethnic composition of the population is one factor because Black people have suicide rates which are about one-half of the rates for White people. This is why the suicide rate in Alabama is smaller than in Vermont. Population density and mean age of population are other factors. In fact, it is precisely for this reason that we analyzed the rate ratio t_m/t_f rather than the rates themselves. Yet even for the ratios the dispersion is too large. We can try to reduce it by considering a sample of states which is more homogeneous. For this purpose we restrict ourselves to states whose density of population is greater than 10 people per square kilometer; to reduce ethnic variability we focus on White people. In addition we restrict ourselves to a specific age interval, namely 20–45 years. All these constraints will of course diminish the number of suicides; to keep it nevertheless at an acceptable level we asked WONDER to sum up the 20 years from 1979 to 1998. As a result of these changes the dispersion of the rate ratio was indeed diminished to $\sigma = 0.45$ down from $\sigma = 0.67$ in the previous case. In spite of this (small) improvement the relationship between r and t_m/t_f remains invisible. If we wish to progress we need to make drastic changes.

9.3.2 Possible options

The main requirement is to increase the interval of variation of the sex ratio r. How can we do that? There are several options.

(1) One may try to focus on groups of immigrants because it is well known that in such groups there are usually more men than women. This should at least enable us to explore the region $r > 1$.

(2) Because men have a shorter life time than women, there are more women than men in populations of elderly people. This may enable us to explore the region $r < 1$.

(3) In some social groups, for instance Roman Catholic priests, there is a high gender imbalance. Can we use such groups as testing grounds?

(4) As some wars provoke a drastic reduction in the male population, the postwar periods should offer a good opportunity to explore the region $r > 1$. For instance, after the War of the Triple Alliance (1864–70) in which Paraguay opposed Argentina, Brazil and Uruguay, the Paraguayan population, which numbered 500,000 before the war, was reduced to about 220,000 of which only 30,000 were males, a situation which corresponds to a gender ratio $r = 0.16$. We come back to this point at the end of the chapter.

Which of the previous options are the most promising? A basic requirement is the *ceteris paribus* condition, by which we mean that with the exception of the gender ratio all other variables should remain unchanged. This requirement is hardly fulfilled in the third option. Indeed, as a social group, Roman Catholic priests differ from the rest of the population in many respects. (i) They may have close connections with other priests, for instance with those with whom they share housing. (ii) They are the result of a process of social selection based on the fact that they decided to become priests. In short, it would be questionable to claim that priests are similar to the rest of the population in all other respects except their gender ratio. With respect to suicide rates, any comparison between priests and the general population would therefore be meaningless. In contrast, the experiment described in option 1 does not have the same problem for in this case we do not have to compare a group of immigrants with the rest of the population; rather we can make comparisons for the same group at different dates, provided that the sex ratio changes in the course of time. For instance, Chinese immigrants in the United States (excluding Hawaii) had a sex ratio of 14 in 1910 and 1.90 in 1950. A possible objection could be raised: are the Chinese immigrants of 1910 really identical to those of 1950? Naturally, it is difficult to answer this question a priori; it is rather an assumption which is justified afterward by observing that the comparison works.

In the next section, we examine more closely the evidence regarding groups of immigrants.

9.4 Suicide rates as a function of sex ratio in groups of immigrants

According to census reports the sex ratio of Chinese immigrants reached a maximum of $M/F = 27$ in 1890 and then decreased steadily, eventually reaching a quasi-equilibrium level of 1.02 in 1970. However, the annual volumes of the *Mortality Statistics of the United States* began to report suicide data for the Chinese population only in 1923.[2] At that time, the sex ratio had already dropped to 5.9. It should be recalled that due to various anti-miscegenation laws, Asian immigrants could not marry white women even if they wished. Moreover, because of restrictive immigration laws, few Asian women could join them. This explains why the sex ratio decreased fairly slowly. In response to this situation, Asian immigrants created a system by which bachelors were adopted as uncles into existing families whose women functioned as surrogate mothers, sisters and aunts to those men (CAPAA 2001). Thanks to the steady decrease of the sex ratio between 1923 and 1970 we have a spectrum of situations characterized by a broad range of sex ratios. This is precisely what we need. Moreover, all these cases are in the region $r > 1$ where the effect is strongest. There is a last requirement that must be checked: are the populations large enough to produce a substantial number of suicides? Table 9.3 summarizes the populations of the three groups under consideration. With a population of only 28,700 the Chinese immigrants in Hawaii are the smallest group. A suicide rate of the order of 30 per 100,000 implies that there are about $30 \times 0.28 = 8$ suicides a year. With such small numbers the variability would be fairly large. This is confirmed by the numbers of suicides observed for

---

[2] High sex ratios among immigrant groups whether in Australia, Saudi Arabia or other countries are fairly common; however, specific suicide data are available in very few cases. Any other documented minority case apart from those that we study in this chapter would be of great interest.

Table 9.3 *Population and sex ratio of immigrant groups in the United States,*
1940

	Continental US		Hawaii		Alaska	
	Population (10^3)	*M/F*	Population (10^3)	*M/F*	Population (10^3)	*M/F*
Chinese	77.5	2.9	28.7	1.3		
Japanese	127	1.3	158	1.2		
Filipinos			52.6	3.5		
Whites					40	2.0

Sources: Census of the United States 1940, Vol. I and Vol. II; Nordyke (1989).

instance from 1936 to 1940, namely 2, 7, 9, 9, 5, which gives a coefficient of variation σ/m as large as 46%. In order to reduce the statistical fluctuations we summed up the suicides over 5-year periods (but 2-year periods in continental US). The choice of this 5-year interval is a compromise between two opposite requirements: reducing the fluctuations but nevertheless keeping enough separate points.

We consider three cases:

- Chinese and Japanese immigrants in continental America, 1923–60.
- Chinese and Japanese immigrants in Hawaii, 1916–50.
- White immigrants in Alaska, 1945–90.

It is natural to consider the immigrants in Hawaii and in the continental part of the United States as two separate cases because the social environments were fairly different. On the continent the Chinese and Japanese communities were in contact with a population which was overwhelmingly of European descent. In contrast, in Hawaii, people of European descent were a small and diverse group. In 1920 it totaled 21% of the population of Hawaii and included many people of Portuguese or Spanish origin who came to Hawaii before it became an American possession. In spite of these different social environments the two cases follow the same rule as far as suicide rates are concerned; this illustrates the robustness of the dependence between suicide rates and sex ratio.

Total suicide rates are shown in Figs. 9.4a, c; male and female rates are shown in Fig. 9.4b for the Japanese population. We did not represent the male and female rates for the Chinese population because the high sex ratio implies that the female population is too small to produce significant numbers of suicides, which results in high statistical fluctuations. Moreover, and for the same reason, the male suicide rate is almost identical to the total suicide rate.

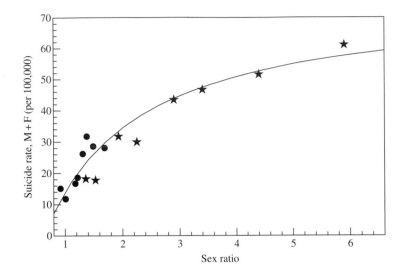

Fig. 9.4a Suicide rates in the Chinese and Japanese communities of the United States. The graph shows the (total) suicide rates of people of Chinese (stars) and Japanese (black dots) descent established in the continental part of the United States as a function of the sex ratio of these communities. The star on the right-hand side corresponds to 1923–4; in subsequent years the sex ratio of the Chinese community decreased steadily. The thin line corresponds to Formula (9.3) with the following parameters: $\bar{t} = 14, \bar{t}_m = 18, \bar{t}_f = 10, k_m = 4.19, k_f = 1.9$ (only $\bar{t}, \bar{t}_m$ and k_m are needed here). *Sources: Mortality Statistics of the United States and Vital Statistics of the United States, various volumes; all volumes are available online on the website of the National Center for Health Statistics.*

In the three graphs of Figs. 9.4a, b, c the thin line represents the suicide rates as defined by the theoretical formulas given earlier. The set of parameters is the same for the three graphs, namely

$$\bar{t}_m = 18 \qquad \bar{t}_f = 10 \qquad k_m = 4.19 \qquad k_f = 1.9 \tag{9.4}$$

The values of k_m and k_f are consistent with the evidence presented in Table 9.1, except for the fact that k_m is somewhat higher than would have been expected. The values of $\bar{t}_m$ and $\bar{t}_f$ are determined by the magnitude of the suicide rates in the vicinity of $r = 1$. Thus, we implicitly assume that these parameters did not substantially change in the course of several decades.

As Fig. 9.4d refers to the White minority of Alaska, it is natural to use another set of parameters, namely

$$\bar{t} = 8 \qquad \bar{t}_m = 21 \qquad k_m = 2.9 \tag{9.5}$$

The value $\bar{t} = 8$ should be compared with the previous value $\bar{t} = (1/2)(\bar{t}_m + \bar{t}_f) = 9$. As can be seen, except for k_m the two sets of parameters (9.4) and (9.5) are not very different.

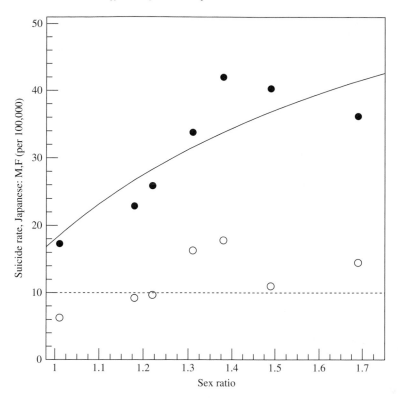

Fig. 9.4b Male and female suicide rates in the Japanese community of the (continental) United States. The filled circles correspond to males, the open circles correspond to females. The data cover the period 1923 (highest sex ratio) to 1955 (lowest sex ratio). The thin lines correspond to Formulas (9.1) and (9.2) with the same set of parameters as in the previous graph. In the region $r > 1$ the theoretical curve for the female suicide rate is a horizontal line. In 1920 the suicide rates of Japanese males and females in Japan were 24 and 15 per 100,000 respectively. *Sources: Same as Fig. 9.4a.*

9.5 Cross-sectional analysis

Our previous attempts in cross-sectional analysis failed because the sex ratio was confined in a narrow interval around $r = 1$. In the light of the previous section one may wonder if the situation can be improved by using data from the early twentieth century, when immigrant groups, particularly in the West, were still characterized by sex ratios notably different from 1. One obstacle for the realization of this plan is the fact that at the beginning of the twentieth century the death registration area comprised only 14 states. Fortunately, the registration area also comprised a number of registration cities in states which were not globally registration states. For instance, in spite of the fact that Nebraska did not belong to the registration area in 1906, its two cities of Lincoln and Omaha were registration

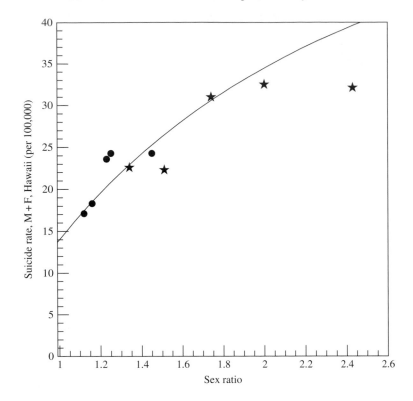

Fig. 9.4c Suicide rates in the Chinese and Japanese communities of Hawaii. This graph is similar to Fig. 9.4a but for the Chinese and Japanese communities in Hawaii. The thin line corresponds to formula (9.3) with the same set of parameters. *Sources: Same as Fig. 9.4a.*

cities. Piecing together these data, we get the graph in Fig. 9.5. There are only five points whose sex ratios are significantly different from 1. In other words, even in this improved form, cross-sectional analysis is less satisfactory than longitudinal analysis.

9.6 Male–female imbalance induced by war

The death of millions of soldiers on the battlefields of World Wars I and II created situations marked by substantial male–female imbalances in several countries. This is illustrated in Figs. 9.6a, b by the cases of France after World War I and Japan after World War II. In contrast to previous cases, the imbalance does not concern the whole population but only specific age groups. In both cases, there are age groups characterized by sex ratios under 0.80.

Under the assumption (that we make for the sake of simplicity) that the average marriage age is approximately the same for men and women, the female age group

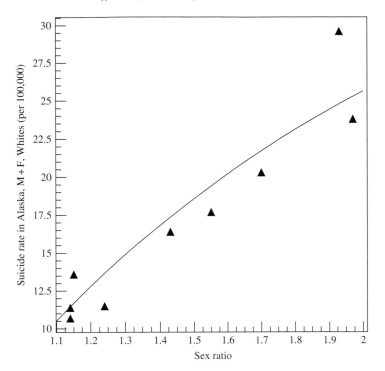

Fig. 9.4d Suicide rates of White people established in Alaska. The graph shows the (total) suicide rates of White people established in Alaska as a function of the sex ratio of this community. The sex ratio decreased steadily in the course of time. The data cover the period 1945 (highest sex ratio) to 1981 (lowest sex ratio). In 1945 the average American suicide rate was 11.5, in 1981 it was equal to 12. Thus, when the sex ratio was equal to 2 the suicide rate of Whites in Alaska was twice the American rate; once the sex ratio had dropped to 1.1 the suicide rate of Whites in Alaska was almost at the same level as in the rest of the United States. The thin line corresponds to Formula (9.3) with $\bar{t} = 8, \bar{t}_m = 21, k_m = 2.9$. *Sources: Same as Fig. 9.4a.*

that should be the most affected by war is the group of women who are of the same age as the soldiers who died on the battlefields. As the latter mainly belong to the group of men who were 20–24 years old during the war, one expects the effect to be largest for women in the same age group; we denote it by $a = (20, 24)$ and will refer to it as the war age group. If we denote by $M(a)$ and $F(a)$ the male and female populations of a, we can apply to these variables the argument used previously. One gets

$$t_f(a)/\bar{t}_f = 1 + \epsilon(k_f - 1) \tag{9.6}$$

where $t_f(a)$ is the female suicide rate in age group a, $\bar{t}_f$ is the female suicide rate in the equilibrium situation $r = 1$ and $\epsilon = 1 - r(a)$ with $r(a)$ denoting the sex

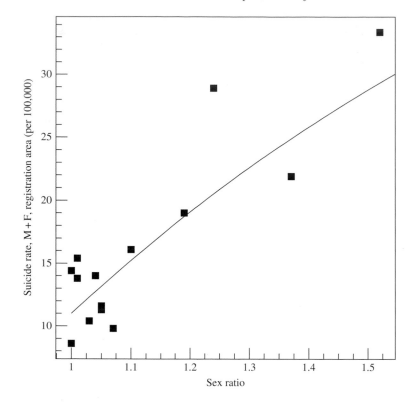

Fig. 9.5 Cross-sectional analysis for suicide rates as a function of sex ratio. The graph is for US states or cities in 1906. Each square corresponds to a state or a city belonging to the death registration area. The places with the highest sex ratio are mainly located in the West. The thin line corresponds to formula (9.3) with parameters $\bar{t} = 11, \bar{t}_m = 27, k_m = 3.7$. No clear conclusion can be drawn because the dispersion is too large. *Sources: Same as Fig. 9.4a.*

ratio in age group a. Formula (9.6) will enable us to estimate the magnitude of this effect in different cases.

France after World War I Figure 9.6a gives $r(a) = 0.80$, for k_f we take the value given in Table 9.1, $k_f = 1.8$; replacing in (9.6) gives $t_f(a)/\bar{t}_f = 1 + 0.2 \times 0.8 = 1.16$.

Paraguay after the Triple Alliance War In this case $\epsilon = 0.84$, if we assume the same value of k_f as in the previous case we get $t_f(a)/\bar{t}_f = 1 + 0.84 \times 0.8 = 1.67$.

This second effect would certainly be large enough to be observable; unfortunately, so far we have not been able to find statistical data for this case. The first effect is fairly small. Moreover, there are several side effects which complicate the detection.

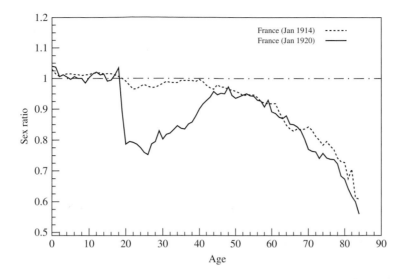

Fig. 9.6a Sex ratio in the French population as a function of age. The graph shows the sex ratio before and after World War I. For the people who are 25 years old in 1920 the sex ratio is 0.78. *Source: Daguet (1995).*

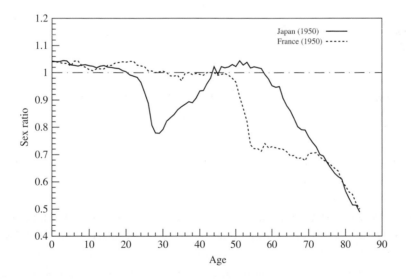

Fig. 9.6b Comparison of the French and Japanese sex ratios in 1950. The trough for French people over 50 corresponds to World War I; with respect to the previous graph the trough became deeper because men have a shorter life expectancy than women. The trough for Japanese people between ages 22 and 42 corresponds to World War II. *Sources: For France, same source as Fig. 9.6a; for Japan, website of the Ministry for Internal Affairs and Communication: http://www.stat.go.jp/english/data/chouki/02.htm.*

(1) We do not know the magnitude of the delay between the war and the occurrence of the excess-suicides. Even though their age group has been depleted, the women of the war age group may entertain the hope of finding a husband in another age group. Therefore their possible suicide may occur several years after the war.

(2) In addition to the male–female imbalance, there may also be female suicides induced by the death of soldiers who were already married or engaged. These suicides may occur soon after the war or perhaps even during the war; in truth, once again we do not know the time lag between link severance and suicide.

(3) In addition to the two previous effects it is likely that the deaths of young males will result in increased suicide rates in the group of their parents. The incidence of the loss of a son has been little studied because this condition is not reported on death certificates. Such an effect should be expected on account of the general rule that the severance of a link results in greater suicide rates.

(4) Still another effect should be mentioned which does not play a role in World War I but may play a great role in World War II. During 1940–3, there were about one million French prisoners of war in Germany. Moreover, in many occupied countries nationals were asked to work in Germany for periods of one or two years. In both cases, the suicides occurring among these populations may not have been recorded in their country of origin. As a result suicide rates may have been underestimated during the time of the war.

In this chapter we discussed the increase in suicide rates due to the impossibility of establishing marital bonds. In the next chapter, we investigate the effects on suicides of cracks that may occur in marital bonds.

Appendix   Suicide rate in a population with a sex ratio $r \neq 1$

In this appendix we establish the formulas for the male, female and total suicide rates. M and F denote the populations of males and females respectively.

First we consider the case where there are more males than females. The F men who are able to get married give rise to a number of suicides equal to $Ft_m(r, \text{married})$ where $t_m(r, \text{married})$ denotes the suicide rate of married males in a population whose sex ratio is r. $t_m(r, \text{married})$ is not necessarily equal to $\bar{t}_m$, the rate in a balanced male/female population because it cannot be excluded that the suicide rate for men is affected by the sex ratio prevailing in the society. However, for the sake of simplicity and because of the lack of empirical data, we will assume that for $r > 1$, $t_m(r, \text{married}) = \bar{t}_m$.

The number of unmarried men is $M - F$. The suicide rate in this subgroup is $k_m t_m(r, \text{married}) = k_m \bar{t}_m$. As a result the total number of male suicides is $s_m = F\bar{t}_m + (M - F)k_m \bar{t}_m$ and the suicide rate $t_m = s_m/M$ becomes

$$t_m = (F/M)\bar{t}_m + (1 - F/M)k_m \bar{t}_m = (1/r)\bar{t}_m + (1 - 1/r)k_m \bar{t}_m$$

If there are more females than males, all males are married and their suicide rate is equal to $t_m(r, \text{married})$; as in the case $r > 1$, we assume that this rate is equal to $\bar{t}_m$. This leads to the result

$$t_m = \begin{cases} \bar{t}_m & \text{if } r \leq 1 \\ (1/r)\bar{t}_m + (1 - 1/r)\,k_m\bar{t}_m & \text{if } r \geq 1 \end{cases} \tag{9A.1}$$

For females, a similar reasoning leads to

$$t_f = \begin{cases} r\bar{t}_f + (1 - r)\,k_f\bar{t}_f & \text{if } r \leq 1 \\ \bar{t}_f & \text{if } r \geq 1 \end{cases} \tag{9A.2}$$

We now use these expressions to compute the total suicide rate t:

$$t = \frac{Mt_m + Ft_f}{M + F} = \frac{rt_m + t_f}{r + 1}$$

Suppose that $r \geq 1$; replacing t_m and t_f by their expressions we get

$$t = \frac{1}{r+1}(\bar{t}_m + \bar{t}_f) + \frac{r-1}{r+1}k_m\bar{t}_m$$

We introduce the total suicide rate $\bar{t}$ in a balanced population: $\bar{t} = (\bar{t}_m + \bar{t}_f)/2$, thus

$$t = \frac{2}{r+1}\bar{t} + \frac{r-1}{r+1}k_m\bar{t}_m \qquad r \geq 1 \tag{9A.3a}$$

A similar calculation in the case $r \leq 1$ leads to

$$t = \frac{2r}{1+r}\bar{t} + \frac{1-r}{1+r}k_f\bar{t}_f \qquad r \leq 1 \tag{9A.3b}$$

Finally, we write the expression for the ratio t_m/t_f which depends only on the ratio $\bar{t}_m/\bar{t}_f$ that we denote by γ:

$$t_m/t_f = \begin{cases} \frac{1}{r + (1-r)k_f}\gamma & \text{if } r \leq 1 \\ [1/r + (1 - 1/r)k_m]\,\gamma & \text{if } r \geq 1 \end{cases} \qquad \gamma = \bar{t}_m/\bar{t}_f \tag{9A.4}$$

Apart from r, which can be estimated from the population data, the ratio t_m/t_f depends only on three parameters: k_m, k_f and γ. The other side of the coin is that (except for large populations) the number of female suicides is fairly small, which results in great fluctuations of t_f and in large error bars for the ratio t_m/t_f.

10

Effect of weakened marital bonds on suicide

In the previous chapter we studied the excess-suicides that arise when a substantial fraction of a population cannot get married. In this and the next chapters we study other mechanisms through which the number of unmarried people is inflated, for instance:

- An increase in the mean age at marriage which results in a greater proportion of unmarried men and women, especially in the 20–30 age group.
- An increase in the rates of divorce or widowhood which results in more people living without a partner.

In sociological studies postponed marriage, divorce and widowhood are usually considered separately.[1] However, seen from a network perspective these are three facets of the same phenomenon, namely the non-establishment or severance of marital bonds. If m, d, w denote the marriage, divorce and widowhood rates respectively, it is possible to define a generalized marriage rate as $m_g = m - d - w$; this rate encapsulates both the establishment of new links and their dissolution through divorce or widowhood. Our objective is to see to what extent these rates are connected with suicide rates. In accordance with the extreme value technique defined in earlier methodological guidelines we first examine three fairly extreme situations: (i) the suicide of young widowers; (ii) the effect of a 200% decline in marriages rates; (iii) the effect of a huge and sudden increase in marriage rate.

10.1 Suicide rate of young widowers

It has been known since Emile Durkheim that widows and widowers have a higher suicide rate than married people. Similar observations have been made in many

---

[1] See for instance Besnard (1997), Smith *et al.* (1988), Luoma and Pearson (2002), Stack (1989, 1990).

time intervals and countries. The following figures provide averages performed over several countries, namely Australia, France, Norway and Switzerland.

Period	Gender	Suicide rate of widowed / Suicide rate of married
1890–1910	Males	2.2
	Females	1.4
1970–1990	Males	3.9
	Females	2.0

Sources: Australian Social Trends 2000, Baudelot and Establet (1985), Besnard (1997), Conrad *et al.* (1911, p. 466), Nizard (1998), Statistiske Centralbyrå (1926, Table 22).

The figures in the table are averages over age groups. They are of the same magnitude as the ratio (suicide rate of non-married people)/(suicide rate of married people) that we considered in the previous chapter. However, if we examine the detailed figures more closely by age groups, a new and fairly surprising effect shows up, namely the fact that the ratio reaches high levels for young widowers. This is documented in Fig. 10.1 which may be compared with Fig. 9.1. This feature has been confirmed by several studies, e.g. Smith *et al.* (1988), Luoma and Pearson (2002). Except for qualitative psychological interpretations, no clear explanation of this effect seems to be known. Nevertheless, and this is the point which is of importance from our perspective, there can be little doubt that the high suicide rate of young widowers is the direct consequence of the severance of the marital bond. Can we conclude from Fig. 10.1 that the strength of the marital bond decreases with age? Before answering, one must examine if the *ceteris paribus* condition is satisfied. It is precisely with this condition in mind that we studied the ratio widowed/married rather than the suicide rate of widowed people taken alone. The numerator and denominator refer to individuals who are identical except with regard to the marital bond. Thus, the natural conclusion seems to be that the strength of the marital bond decreases with age. However, the fact that the suicide rate of young widowers is also much higher than the rate for bachelors suggests that the loss of a wife has effects which go beyond the sphere of the family unit.

In a previous chapter we argued that molecular interactions often provide an insightful parallel to less known social interactions. What does this parallel tell us in the present case? What would be the physical analogue of the inability to establish marital bonds, for instance because of a male–female imbalance? It is similar to what occurs in a water–ethanol mixture when there are less than three or four water molecules for each ethanol molecule; indeed, the normal configuration of

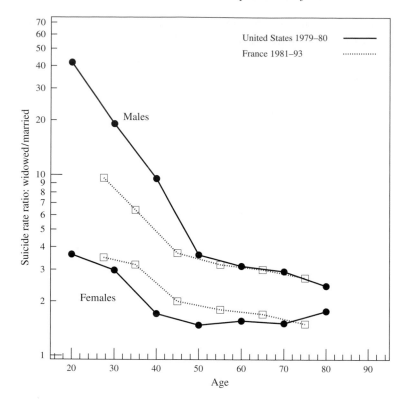

Fig. 10.1 Ratio of suicide rates of widowed and married people. The age groups are 15–24, 25–34, 35–44, etc. in the United States and 25–9, 30–9, 40–9, etc. in France. Young widowers have a very high suicide rate, of the order of 1,000 per 100,000. This rate is higher than the rate for married people but it is also higher than the rate for unmarried people. As the number of young widowers is small, the number of suicides in the youngest age groups is not large; in the United States over 1979–80 there are 34 suicides in the 15–24 age group and 112 in the 25–34 age group. Consequently, these points have fairly large error bars due to statistical fluctuations. The fact that young widowers have very high suicide rates is confirmed by several studies, e.g. Smith *et al.* (1988), Luoma and Pearson (2002). *Sources: United States, suicides: Vital Statistics of the United States, 1979, 1980, Vol. II, Part A, p. 323; population of age groups by marital status: Statistical Abstract of the United States (1981, p. 38); France: Besnard (1997).*

a mixture of water and ethanol comprises water–ethanol complexes in which one molecule of ethanol is surrounded by three, four or even more molecules of water. Thus, if the concentration of ethanol is too high the excess ethanol molecules remain unbonded to water molecules; instead of experiencing the (higher) inter-action of the mixture they experience only the interaction they would have in pure ethanol. The analogue of the ratio (never married)/married would be the interaction ratio (ethanol not bonded to water)/(ethanol bonded to water).

It is more difficult to find a physical parallel to the notion of widower. In physics, the standard way to define the binding energy of an interaction A–B is to estimate the cost in energy to separate A and B until the distance d_{AB} becomes infinite. The closest analogue of this definition is the tie severance brought about by divorce or widowhood. However, once the molecules are apart, they do not have any "memory" of their former situation; in other words, while the molecular parallel explains fairly well why the suicide rate for divorced or widowed people is approximately equal to the rate for unmarried people, it does *not* explain why the suicide rate for young widowers is so much higher than the suicide rate for bachelors of the same age. One way to resolve this difficulty would be to say that for young husbands the loss of their wives also weakens their ties with the rest of society.

10.2 Effect of falling marriage rates

At the beginning of the chapter we introduced the notion of generalized marriage rate, m_g, defined as $m_g = m - d - w$. The widowhood rate w may be of significance for young generations in time of war, but in normal times it mainly concerns age groups over the age of 60; as it is fairly constant in the short and medium term it will be left aside in most of the investigations conducted in this chapter. In 1910, the US marriage and divorce rates were equal to 10 and 1.0 per 1,000 population respectively. In other words divorce played a fairly small role; the situation remained the same until the 1970s. In 1965 the divorce rate was still only about 20% of the marriage rate, but in 1975, marriage and divorce rates were equal to 10.0 and 5.0 respectively. In other words, after 1975 the variations in d significantly affected m_g.

In France, the divorce rate also picked up in the 1970s but in addition the marriage rate fell considerably. In 1970 the marriage and divorce rates were equal to 7.8 and 0.8 per 1,000 population respectively; in 1995, they were equal to 4.9 and 2.1 (Eurostat 2004, pp. 119, 126). As a result m_g dropped from 7.0 to 2.8. What was the effect on suicide rates? Figure 10.2 shows that the changes in suicide and effective marriage rates are fairly parallel. To make the graph easier to read we represented $-m_g$ rather than m_g. Table 10.1 shows that the correlation between the suicide rate s and m_g is equal to -0.95. Naturally, one should keep in mind the possibility that the two changes are in fact due to a third factor. A further test is to look at the time lag between the two series. If the changes in m_g really bring about the changes in s, they must also precede them. In principle this argument is correct, but one must observe that what really matters is the moment when the bonds are established or broken. Two persons can become engaged months or even years before they get married. In the 1970s it became

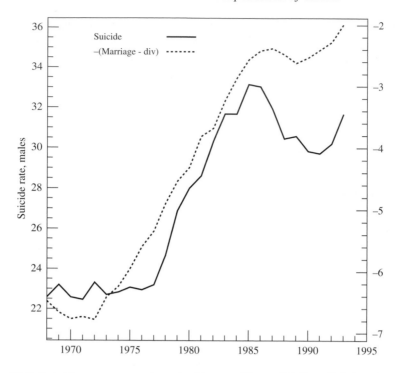

Fig. 10.2 Suicide versus marriage in France. The solid line (left-hand scale) refers to the male suicide rate per 100,000 males. The difference between marriage and divorce rates can be seen as a net (or effective) marriage rate; in order to make the graph more readable, the broken curve represents the opposite of the net marriage rate (right-hand scale). Due to this change, the two curves are almost parallel instead of being anti-parallel; their coefficient of correlation is 0.95. The correlation between m_g and suicide rates is 0.95. *Sources: Besnard (1997, p. 757), Eurostat: Statistiques de population, Chapitre G: Nuptialité (2004, pp. 119, 126).*

more and more common especially in European countries for two partners to live as husband and wife without being married and to get married only after the birth of the first child. Through this mechanism the delay between bond genesis and marriage could reach two or three years.[2] Similarly, the marital bond is often broken months or years before divorce officially occurs.

On average during the period 1970–95 the marriage rate is 6 times higher than the divorce rate (in 1970 the ratio is 9.7, in 1995 it is down to 2.3) which means that marriage deficit is the dominant factor. The main difference between postponed marriage and divorce is that the first affects young age groups while the second affects middle age groups. As an additional test one may therefore

---

[2] Figure 10.2 shows that in the early 1970s, m_g precedes s by several years but in the early 1990s the time lag becomes much shorter.

Table 10.1 *Correlation between marriage and suicide of males in*
France, 1968–1993

x	Correlation (x, y)	a	b
Mean marriage age	0.80	2.5 ± 0.7	-36 ± 1.0
Marriage rate	−0.96	-3.0 ± 0.4	44 ± 0.4
Marriage rate − divorce rate	−0.95	-2.2 ± 0.3	37 ± 0.5

Notes: y is the suicide rate of males, a, b denote the coefficients of linear regression in $y = ax + b$.
Source: Besnard (1997, p. 757).

check whether the excess-suicides are indeed more correlated with the young age groups. Besnard (1997, p. 748) provides the following correlations between suicide and (first) marriage rates as a function of age group.

	20–29	30–39	40–49	50–59
Males, 1968–93	−0.77	−0.88	−0.63	0.21

Thus, the maximum of the correlation indeed occurs in young age groups rather than in groups aged over 40. The fact that the maximum occurs in the 30–39 age group can be related to the observation (already made in Fig. 9.1) that remaining unmarried seems more difficult to accept after the age of 30, perhaps because in this age group the number of unmarried people is lowest.

10.3 Effect of a sudden upsurge in marriages

Sudden upsurges of marriages may occur at the end of wars. For instance, the US marriage rate jumped from 11 per 1,000 in 1944 to 16 in 1946. An even bigger upsurge occurred in Sweden in December 1989: there were 64,218 marriages in this month, a number which is about 26 times larger than the average for other December months in the period 1985–93. What was the effect on suicides? Before answering this question one may wish to know what caused this huge increase. It was due to a regulation through which women born before 1944 were eligible for a pension after the death of their husbands provided they were married at the end of 1989; in addition the husband had to be under 60 at the time of the marriage. Did these marriages really correspond to the genesis of new bonds or were they merely validations of existing unions?[3] It is difficult to know. However, Fig. 10.3 shows that the effect on suicides was not very impressive, which seems to suggest that indeed most of the marriages were simply validations of existing unions. This

---

[3] In addition, some of the marriages may have been marriages of convenience not corresponding to real bonds.

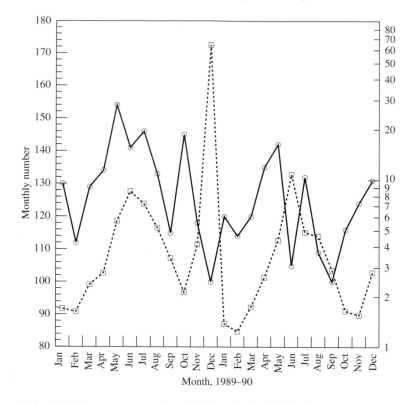

Fig. 10.3 Suicides versus marriages in Sweden in 1989–90. Solid line: monthly number of suicides in the two years 1989 and 1990 (left-hand scale); broken line: monthly number of marriages (right-hand scale in thousands of marriages). December 1989 was marked by an exceptionally high number of marriages; the fact that this month is a minimum in the number of suicides is not due to the seasonal pattern of suicides for it can be verified that the deseasonalized suicide series also displays a minimum in December 1989. However, on account of the huge number of marriages, one would have expected a more pronounced minimum. Probably most of these marriages were validations of already existing unions. *Sources: Marriages: website of Eurostat; suicides: data communicated by Dan Bernhardson from the Epidemiological Center of Sweden to whom I express my gratitude.*

outcome makes this case rather disappointing but on the other hand it also attracts our attention to the distinction between the bonds themselves and the way they are registered. Great caution must be exercised in this respect.

10.4 Connection between mean age of marriage and suicide rate

The theoretical argument used in the previous chapter can fairly well be adapted to the present situation. Instead of reasoning in terms of marriage rates it is simpler to reason in terms of mean age of marriage. Thus a first step is to show that the two variables are almost equivalent. Besnard (1997) reports that in France between 1968 and 1993 the correlation between mean age of marriage and marriage rate was −0.89 for males and −0.88 for females. The regression is as follows:

$$x = \text{Marriage rate per 1,000 population}, \quad y = \text{Mean age at first marriage},$$
$$y = ax + b$$

	a	b
Males	-0.90 ± 0.17	30.8 ± 0.2
Females	-0.92 ± 0.18	28.8 ± 0.2

Similar relationships hold in other countries as well. For instance, in Australia during 1977–2002 the correlation is −0.93 (for both males and females); in the Netherlands between 1960 and 2004, it is −0.70; in Sweden in the same time interval it is −0.79. These results lead us to the conclusion that marriage rate (x) and mean age of marriage (y) are related by a relationship of the form $y = ax + b$, where a is about −1 and b is of the order of 30 years.

The argument has the same starting point as the reasoning used in the previous chapter, namely the fact that the suicide rates of non-married persons are about twice as high as those of married persons. Suppose for instance that between time t_1 and time t_2 the mean age of marriage increases from 23 to 28 years. Let us denote by g_m the age group 21–30. Initially, only 3/10 of the age group are unmarried, but at time t_2 this proportion jumps to 8/10; therefore one expects an increase in the suicide rate of this age group given by

$$t_{28}(g_m) - t_{23}(g_m) = [0.8k\bar{t} + 0.2\bar{t}] - [0.3k\bar{t} + 0.7\bar{t}] = 0.5(k-1)\bar{t}$$

where $\bar{t}$ and $k\bar{t}$ are the suicide rates of married and unmarried people respectively. More generally, if the mean age of marriage increases from a_1 to a_2 the change in suicide rate will be

$$t_{a_2}(g_m) - t_{a_1}(g_m) = [(a_2 - a_1)/10](k-1)\bar{t}$$

In the next sections we design several experiments to test this kind of relationship.

10.5 Longitudinal test

As our objective is to observe the effect of a change in marriage rates on suicides, we must select time intervals which display substantial variations in marriage rate. In many industrialized countries there was a notable trough in marriage rates during the time interval 1920–40. This is illustrated for the case of the United States in Fig. 10.4. In order to analyze the response of the suicide variable to such a change,

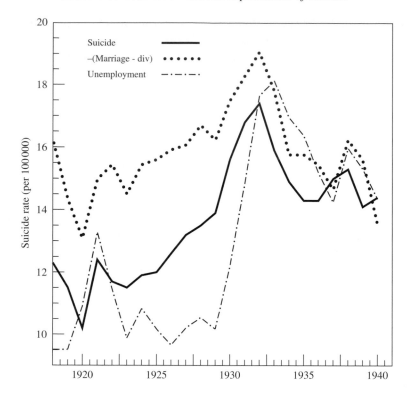

Fig. 10.4 Suicide rate versus marriage rate and unemployment in the United
States. The left-hand scale refers only to suicide rates (the scales for the two other
curves are not displayed). The graph has been made more readable by replacing
the effective marriage rate by its opposite. It can be noted that the suicide rate
was increasing during the whole decade 1920–9 whereas the unemployment
rate increased only after 1929. The correlations are given in Table 10.2; it is
interesting to note that the slope of the regression line in the upgoing phase is
twice as high as in the downgoing phase. *Sources: Suicide: Historical Statistics
of the United States (1975, p. 64); Marriage and divorce: Vital Statistics of the
United States 1949, Part I, p. LIX–LXII.*

we can use longitudinal or cross-sectional analysis. In longitudinal (also called tem-
poral) analysis we estimate the correlation between m_g and the suicide rate s in
a given country in the course of time. This is the purpose of the present section.
In the cross-sectional (or spatial) analysis, one studies how different countries or
regions respond to a similar input signal; this will be examined in the next section.

We now come back to Fig. 10.4; it displays three variables: marriage rate,
suicide rate and unemployment. The suicide rate is seen to peak in 1932; often
in the literature this peak is attributed to the Great Depression but this is a poor
explanation because the suicide rate started to climb in 1920 whereas the Great
Depression began in 1929, as is reflected in the curve of the unemployment rate.

Table 10.2 *Relationship between suicide, marriage and unemployment rates,*
United States, 1918–1940

x	Correlation (x, y)	a	b
Phase of increase in suicide rate			
(1918–32)			
Marriage rate	−0.94	−1.9 ± 0.6	32 ± 0.4
Marriage rate − divorce rate	−0.95	−2.0 ± 0.4	30 ± 0.3
Unemployment rate (%)	0.76	0.2 ± 0.1	11 ± 0.7
Phase of decrease in suicide rate			
(1933–40)			
Marriage rate	−0.66	−0.42 ± 0.4	19 ± 0.3
Marriage rate − divorce rate	−0.64	−0.51 ± 0.5	19 ± 0.4
Unemployment rate (%)	0.60	0.1 ± 0.1	13 ± 0.4

Notes: y is the suicide rate, a, b denote the coefficients of linear regression in $y = ax + b$. The correlations between suicide and unemployment are smaller than between suicide and marriage rates.
Sources: see Fig. 10.4.

The curve of the suicide rate and of m_g are highly correlated: the correlation is −0.95 (Table 10.2) whereas the correlation between unemployment and suicide is substantially lower.

We do not yet know what brought about the fall in marriage rates after 1920. Naturally, many tentative explanations could be proposed,[4] but our main objective is to understand the connection between changes in the strength of ties and suicide. What determines that strength is beyond the limits of our present study.

The next section complements the longitudinal study by cross-sectional analysis.

10.6 Cross-sectional analysis

As we have already done in the previous chapter we perform the cross-sectional analysis at the level of the American states. For some reason, marriage rates at state level were not collected between 1933 and 1939. In what follows we consider two time intervals: 1925–32, which was characterized by a fall in marriage rate, and 1932–40, which was marked by an increase in marriage rates. In any

---

[4] For instance, it may be argued that in some sectors the crisis began much earlier than October 1929. For instance, real estate prices fell after 1925, which provoked a crisis in the housing sector, particularly in New York and Florida. In the farming sector the crisis actually began in the early 1920s. In 1909 the price of a bushel of wheat (Hard Winter No. 2 in Kansas City) was $1.10; in 1924 the price was almost the same but in the meanwhile the consumer price index had almost doubled (*Monographies de produit* 1950, graph 6). As a result of this difficult situation American farmers ran into debt, a debt which became intolerable after 1929 when real wheat prices fell even lower.

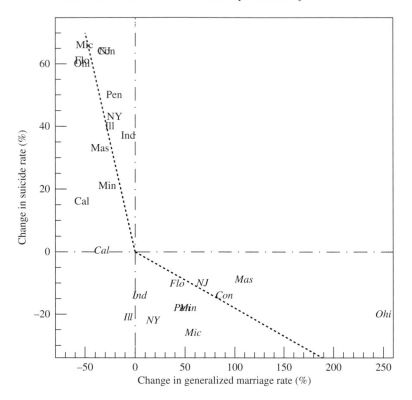

Fig. 10.5 Relationship between marriage rates and suicide rates in 12 American states. The states are the following: California, Connecticut, Florida, Illinois, Indiana, Massachusetts, Michigan, Minnesota, New Jersey, New York, Ohio, Pennsylvania. The changes are computed over two different time intervals: 1925–32 and 1932–40 (labels in italic). The plot shows that the relationship between marriage and suicide is weaker in the second period than in the first, an observation already made in the longitudinal analysis of Fig. 10.4. *Sources: Marriage, divorce: Statistical Abstract of the United States 1928, 1934, 1953. Suicide: Linder and Grove (1947), Mortality Statistics (1932).*

cross-sectional analysis a central question is the selection of the regions that will be considered. Naturally, this must be done on the basis of objective criteria. The simplest choice would be to take all states, but for a number of reasons this is neither practicable nor suitable. (i) In 1960 the marriage registration area comprised only 33 states (see *Vital Statistics of the United States* Vol. III, Section 1, p. 12), which means that for the other states the data were estimates and therefore not completely reliable. (ii) In some of the registration states there was a high proportion of marriages between non-residents. For instance, in 45% of the marriages which took place in Maryland neither the bride nor the groom was a resident of Maryland. Such preferences for some states were probably related to differences in state legislation; Las Vegas in Nevada provides an illustration of

Table 10.3 *Cross-sectional analysis of the relationship between changes in suicide and marriage rates, 1925–1940*

Sample	Number of territories	Correlation	a	b
American states	12	−0.66	−0.32±0.15	19±10
Industrialized countries	10	−0.46	−0.34±0.3	6±5

Notes: The table gives the correlation and regression coefficients between percentage changes of effective marriage rates (m_g) (i.e. marriage rate − divorce rate) and suicide rates. For each territory we consider changes over two time intervals I_1 and I_2; over the first interval marriages decreased in most territories whereas during the second they mostly increased. The negative correlations show that suicide rates increase (decrease) when m_g decreases (increases). Due to data availability I_1 and I_2 are not exactly the same for the two samples; for American states: $I_1 = (1925, 1932)$, $I_2 = (1932, 1940)$, for industrialized countries $I_1 = (1927, 1932)$, $I_2 = (1932, 1937)$. The 12 American states are listed in the caption of Fig. 10.4; the 10 countries are the following: Australia, Belgium, Canada, Denmark, Finland, Japan, Netherlands, Sweden, Switzerland, United States. It should be noted that the impact of marriage rates is stronger in the region of negative m_g changes than in the region of positive changes.
Sources: American states: see Fig. 10.4. Industrialized countries: marriages, divorces, suicides after 1931: *Annuaire Statistique de la France, Résumé Rétrospectif* (1966, p. 9*, 10*, 35*); suicides before 1931: *Epidemiological and Vital Statistics Report*, World Health Organization, Vol. 9, 1956, Geneva.

this kind of situation. (iii) Afro-Americans have lower suicide rates than Whites; including states with a high proportion of Afro-Americans might therefore distort the results. (iv) Finally, there is little benefit in including states with small populations (say less than one million) because they will have only few suicides, which will result in high statistical fluctuations. Including such states would increase the level of noise beyond necessity.

Eventually, 12 states turned out to satisfy these requirements. The corresponding changes in effective marriage and suicide rates are shown in Fig. 10.5. The correlation and regression coefficients are given in Table 10.3. From the plot it is clear that the relationship is fairly different depending on whether the changes in m_g rates are positive or negative. Obviously, decreases in m_g rates have a greater impact on suicides than increases. One gets very similar results for a sample of 10 industrialized countries.

Other time intervals In this section we focused on the time interval 1920–40. It is of course not the only episode for which a change in the number of suicides can be explained by a change in the number of marriages. At the beginning of the chapter we examined the case of France in 1968–93. Other possible episodes are Japan in 1905–18 or in 1946–65. In a more general way, on what criteria should one rely to select time intervals which are appropriate for this kind of study?

- The first requirement, as already mentioned, is that marriage rates change substantially.
- The second requirement is that no other effect comes into competition with the influence of marriage rates. In the next chapter we will see that immigration can have a marked effect on suicide rates, especially if the immigrants come from countries where suicide rates are much higher (or much lower) than in the country of destination. Therefore, if one wishes to focus on the impact of marriage rates, one must select periods during which immigration is small. In most countries the inter-war period 1918–39 was an excellent time in this respect.

Although it is the most important, the marital bond is of course not the only link between an individual and the rest of the society. In the next chapter we examine situations in which individuals are deprived not only of marital bonds but also of links with relatives or friends. In particular we will try to estimate how long it takes to rebuild a network of links when individuals have to switch from one social environment to another.

11

Effect of social isolation on suicide

In the previous chapter we studied the effect of weakened marital bonds on suicide. Of course, the family unit is only one of the entities to which an individual belongs. In the language of physical interactions, family bonds are short-range ties whereas links with society at large can be seen as long-range ties. Intermediate between short and long are medium-range ties with relatives, friends, neighbors, colleagues, etc. Typically, these links have a duration of a few years, whereas family ties have a duration of several decades.

A quick estimate of the strength of long-range ties can be obtained through the extreme value approach. It leads us to analyze the impact on suicides of major events such as the attack on Pearl Harbor and September 11, 2001. We will see that even such dramatic events have no visible effect on suicide rates. In the subsequent sections we consider various situations of social isolation such as illness, imprisonment or emigration. In each case the excess-suicides with respect to the general population can be taken as a proxy of the social ties which are missing or severed.

11.1 Effect of major historical events on suicide

First we examine the effect of the attack of September 11; then we discuss more briefly the consequences of the attack on Pear Harbor.

11.1.1 September 11, 2001

In accordance with the extreme value approach we consider the number of suicides in the place where one may expect the effect to be greatest, namely in New York City. The series in Table 11.1a gives the monthly numbers of suicides in the three years 2000–2. At first sight, the figure for September 2001 does not appear abnormally lower or higher, but the matter needs to be considered

Table 11.1a *Monthly number of suicides in New York City, 2000–2002*

Year	Jan.	Feb.	Mar.	Apr.	May	June	July	Aug.	Sep.	Oct.	Nov.	Dec.
2000	44	36	33	48	44	37	27	43	32	39	35	30
2001	32	24	32	46	48	42	51	44	33	33	34	43
2002	42	35	36	48	56	34	42	36	44	40	50	32

Source: New York City Department of Health. Many thanks to Mr. J. F. Kennedy for sending me these data.

more carefully, in particular because of the seasonal pattern of suicide rates. The monthly means of 2000 and 2002, $e_m = [s_m(2000) + s_m(2002)]/2$, $m = 1, \ldots, 12$, were taken as reflecting the monthly pattern; thus, the relative differences $p_m = [s_m(2001) - e_m]/e_m$ furnish estimates of how much the monthly data of 2001 differ from the normal monthly pattern. This variable is displayed in Fig. 11.1a. The graph confirms our first conclusion, namely that the fluctuation of September 2001 is smaller than the standard deviation of monthly fluctuations.

Incidentally, it can be observed that this episode provides an objection to the Werther effect that we discussed in Chapter 3. In its initial presentation by Phillips (1974) the Werther effect is presumed to be an imitation effect through which people who read or hear about recent occurrences of suicides are tempted to commit suicide themselves. On September 11, people committing suicide by jumping from the Twin Towers were shown repeatedly on TV, yet without producing any identifiable imitation effect.[1]

As the attack of September 11 was an event of short duration, a kind of delta function shock, its effect on suicides may also be short-lived. In other words, one may wonder if there is a visible effect on daily numbers of suicides. With only about 40 suicides per month in New York City, there would be less than two suicides a day; clearly, a series with such small numbers would be meaningless. In other words, daily suicide data will be of significance only at the level of the whole country. Assuming an annual suicide rate of 10 per 100,000 and a population of about 300 million we expect a daily number of suicides of the order of $30,000/365 = 82$. This is indeed the correct order of magnitude as shown by the data in Table 11.1b. The table gives the data for the first two weeks of September starting with the first Sunday of September. Thus,

---

[1] It could of course be argued that there have been two opposite effects which canceled each other out: (i) an imitation effect which led to additional suicides; (ii) a "polarization effect" that resulted in a reduction of suicides. So far, however, there is no evidence in favor of such an interpretation.

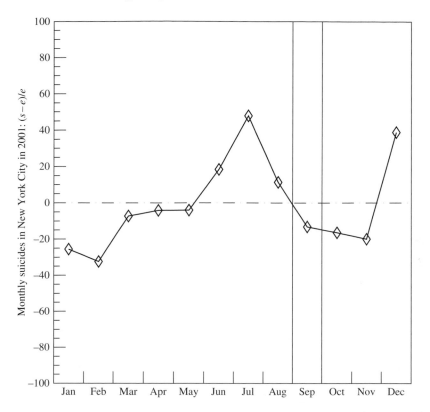

Fig. 11.1a Effect of the attack of September 11, 2001 on monthly suicides in New York City. The curve represents the relative percentage differences $p_m = [s_m(2001) - e_m]/e_m$, $m = 1, \ldots, 12$, between the numbers of suicides in 2001 and the means, e_m, for the corresponding months in 2000 and 2002. The purpose of this transformation is to discard the influence of the seasonal pattern of suicides. The small negative fluctuation of p_m in September 2001 is well within the bounds of the standard deviation $\sigma_p = 25\%$. *Source: See Table 11.1a.*

the figures in the first column are for Sundays, those in the second column are for Mondays, etc. September 11, a Tuesday, is the day delimited by asterisks. At first sight this figure is neither substantially lower nor higher than other Tuesdays. In order to get rid of the daily pattern we use the same procedure as previously, that is we compute the daily means e_d of 2000 and 2002; the relative differences $[(s_d(2001) - e_d)]/e_d$ then give an estimate of how much the daily data differ from the daily pattern. This variable is displayed in Fig. 11.1b. The graph confirms our first impression in the sense that the fluctuation of September 11 was of a size which occurs every two weeks as a result of random fluctuations.

Table 11.1b *Daily number of suicides in the United States during the first two weeks of September, 2000–2002*

Year	S	M	T	W	T	F	S	S	M	T	W	T	F	S
2000	73	70	88	85	80	78	77	68	85	100	99	82	78	74
2001	73	80	88	107	77	75	87	77	89	*73*	86	89	82	72
2002	83	87	110	102	79	96	59	72	103	81	80	87	90	75

Notes: The table gives the number of suicides for the first two weeks of September starting from the first Sunday in the month. September 11, 2001 corresponds to *73*.
Source: National Health Center Website (Vital Statistics, Mortality, Table 304).

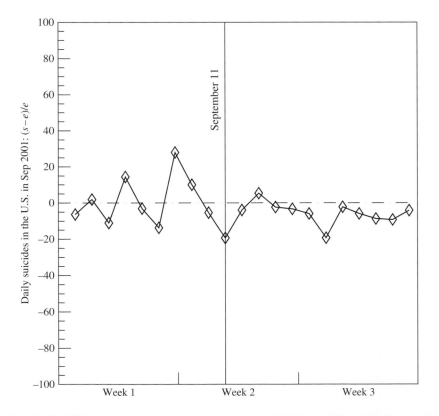

Fig. 11.1b Effect of the attack of September 11, 2001 on daily suicides in the United States. Week 1 begins on the first Sunday of September. A transformation $p_d = [s_d(2001) - e_d]/e_d, d = 1, \ldots, 21$, was performed to get rid of the daily pattern of suicides; e_d is the mean number of suicides on day d for 2000 and 2002. The negative fluctuation on September 11, 2001 is 1.78 times the standard deviation. The likelihood of such a fluctuation to occur randomly is one chance in 13; indeed the graph shows that there are two fluctuations of that magnitude (or higher) in the 21-day period. *Source: See Table 11.1a.*

11.1.2 Pearl Harbor

Was the suicide rate in December 1941 abnormal (either lower or higher) in some way? By applying the same procedure as previously we find that the change in $p = (s - e)/e$ is 1%, well below the standard deviation of the series which is approximately equal to 5%. Taking a broader view, we can observe that the suicide rates in the United States decreased from 1933 to 1944 and that the beginning of the war in December 1941 had little effect on this trend. For instance, the maximum of 1942 (which occurred in April) is at the same level as the maximum of 1941 (also in April).

11.1.3 Weakness of long-range ties

In conclusion, major events which occur at the level of the nation seem to have no effect whatsoever on suicide rates. In other word, the long-range links between individuals and the nation appear to be much weaker than short-range links. This applies to events of short duration such as September 11 as well as to events of longer duration such as Pearl Harbor and the subsequent battles which took place in the Pacific in the spring of 1942. It may be observed that these events affected *directly* the life of only a small proportion of the people. In both cases the number of victims was about 3,000, which represents a small percentage with respect to the population of the United States but also with respect to the population of New York City (0.4%). Naturally, the situation changed completely after 1942 when the lives of millions of Americans were *directly* affected by the war, in particular for those who served in the armed forces. At this point one would certainly expect a substantial impact on suicide rates because such changes alter family ties. However, in such a period of time, many social changes took place simultaneously which means that it is not a favorable situation for sorting out the effects of different factors.

The picture which emerges from this discussion is that of a spectrum of bonds and links with strong short-range marital ties at one end and weak long-range interactions at the other end. In between are medium-range interactions for which we have no clear evidence so far. It is the purpose of the next sections to shed some light on this question.

11.2 Effect of social isolation on suicide

In what follows we consider several situations which are characterized by the severing of one or several sorts of ties. Table 11.2 provides an overall view of the corresponding suicide rates.

Table 11.2 *Effect of social isolation on suicide*

Condition	Suicide rate (s) (per 100,000)	s/(rate in general population)
People with schizophrenia	200	20
Inmates in solitary confinement (United States)	140	7
Inmates in prison (Britain)	110	5
German immigrants in the United States (1870)	40	2

Notes: The purpose of the table is to give overall orders of magnitude. In the first case we assume a suicide rate in the general population of 10 per 10^5 which is approximately the rate for the US or British population. In the second and third cases the general population is the male population. In the last case "general population" refers to the German population in 1870 which had a suicide rate of about $20/10^5$.
Sources: Schizophrenia: Schizophrenia Society of Ontario (http://www.schizophrenia.on.ca); for the other cases, see subsequent tables.

Why did we include people with schizophrenia in a table about social isolation? Schizophrenia is a severe mental illness characterized by a variety of symptoms among which the loss of contact with reality and social withdrawal play a great role. People with schizophrenia avoid others or act as though others do not exist; for instance, they may lack interest in participating in group activities. Clearly, the interpersonal links of people with schizophrenia are severely curtailed. Suicide rates of people with schizophrenia are not well known. One reason may be the fact that the characterization of schizophrenia involves a degree of uncertainty. Estimates range from 10 to 20 times the rate in the general population.

Asperger's syndrome is a form of autism, but what characterizes people with this syndrome compared with others with autism is that they have good language skills and average or above average intelligence quotient. Instead of being withdrawn, a person with Asperger's syndrome may talk on and on, regardless of the listener's interest. The average suicide rate of people with Asperger's syndrome is not known precisely but a study by the National Autistic Society in Britain found an 8% rate of suicide and attempted suicide. If confirmed this would be a very high rate, for it is usually estimated that in the general population there are 10 suicide attempts (with hospitalization) for every suicide. This would put the rate of suicide and attempted suicide at about 0.15%, which is 50 times smaller than the above figure of 8%. People with Asperger's syndrome also seem to have a higher divorce rate than the general population.

The other cases in Table 11.2 concern suicide among inmates and immigrants. In contrast to the previous categories, these are well-defined situations for which detailed data are available.

11.2.1 Suicide among inmates

Persons who are arrested and jailed are removed from their normal social environment and, as a result, their short- and medium-range links are suddenly severed. However, after several months in jail, inmates are likely to build up new ties, for instance with other inmates or with guardians, chaplains, lawyers. One would expect therefore that it is during the first days in jail that the disaggregation of social ties is the most severely felt. This prediction is indeed confirmed by observation. It turns out that suicide rates are particularly high during the first days spent in jail (Fig. 11.2). A study of jail suicide in the United States performed in 1986 found that during the first hours after incarceration the suicide rate reaches very high levels of the order of 5,000 per 100,000 (Fig. 11.2). This study found that 51% of the suicides which occur in lockup jail happen in the first 24 hours after the arrest. Lockup jails are detention facilities where detainees usually stay

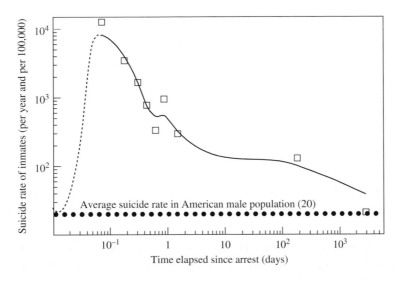

Fig. 11.2 Suicide rate of inmates as a function of the time elapsed since their arrest. In the two days following incarceration, the suicide rate decreases rapidly. Subsequently, the decrease continues at a much slower rate. After two years the suicide rate becomes almost identical to the rate in the general male population. The decrease roughly follows a power law (at least until the plateau phase) with an exponent equal to -1.1: suicide rate = 1/(time elapsed since incarceration)$^{1.1}$. *Sources: Hayes and Rowan (1988, p. 36); Roehner (2005a, pp. 669–70).*

for less than 72 hours before being transferred to county jails; in US terminology, "jails" designates facilities for periods of imprisonment of less than one year whereas prisons keep detainees for periods exceeding one year.

How accurate and reliable are these data? An official report (Hayes and Rowan 1988) found under-reporting of suicides to be of the order of 40% nationally but with great differences between states. Thus, in New York State no under-reporting was identified whereas in Alabama, Louisiana, Pennsylvania and Tennessee under-reporting was as high as 50%. As there is no mention of over-reporting we can be assured that the statistical data provide at least reliable lower bounds.

Table 11.3 gives some data for other countries. These statistics do not distinguish between short- and long-term facilities. Most of the figures are between 100 and 200, which is consistent with the rates observed in US county jails.

It could be argued that the high suicide rate shortly after incarceration should be attributed to the trauma of the arrest rather than to the severance of social links. Possible ways of choosing between the trauma and the link interpretation would be to find situations in which there is a sharp increase in suicides without initial trauma or, conversely, which involve an initial trauma but do not result in higher suicide rates. In what follows both tests will be used.

11.2.2 Suicide in solitary confinement

The evidence on which we rely in this section concerns suicides of inmates in solitary confinement cells in New York State prisons. Also called "special housing units" these cells are used for the purpose of punishing inmates for rebellious behavior. The regime of these cells implies confinement 23 hours a day. On December 31, 2001 there were 3,654 inmates in solitary confinement representing 5.4% of the total population of the 67,000 inmates in New York State.

Over the four years 1998–2001 there were on average 5 suicides per year in solitary confinement cells, which results in an annual rate of $5/(3,654/10^5) = 137$ per 10^5; this figure should be compared with the rate of 26 per 10^5 in prison (*Poughkeepsie Journal*, articles by Mary B. Pfeiffer, December 16, 2001 and April 14, 2002). As the numbers of suicides in confinement are small, one may wonder to what extent the difference $d = 137 - 26$ is really significant. Through a reasoning for which we omit the technical details one arrives at the conclusion that $d/\sigma(d) = 1.9$. In other words (see Table 3.2) the likelihood that the difference d is due to a random fluctuation is 1 chance in 40.

Having assessed the significance of this difference in suicide rates, let us now examine how it can be explained in the network perspective on the one hand and in the trauma interpretation on the other hand. In the first perspective the higher suicide rate would be attributed to the severance of ties with inmates

Table 11.3 *Suicide rates among inmates*

	Type of institution	Time elapsed since incarceration (T)	Location	Time interval	Annual suicide rate (per 100,000)
1	Lockup	$T < 72$ hours	New York State	1990–1999	900
2	Lockup	$T < 72$ hours	South Dakota	1984	2,975
3	Jail	72 hours $< T <$ 1 year	Texas	1981	137
4	Jail	72 hours $< T <$ 1 year	South Carolina	1984	166
5	Jail	72 hours $< T <$ 1 year	United States	1986	107
6	Jail	72 hours $< T <$ 1 year	New York State	1986–1987	112
7	Prison	1 year $< T$	United States	1984–1993	21
8	Not specified	Not specified	Belgium	1872	190
9	Not specified	Not specified	England	1872	112
10	Not specified	Not specified	Saxony	1872	860
11	Not specified	Not specified	Canada	1984–1992	125
12	Not specified	Not specified	New Zealand	1988–2002	123
13	Not specified	Not specified	England	1990–2000	112
14	Not specified	Not specified	France	1991–1992	158
15	Not specified	Not specified	Australia	1997–1999	175
16	Not specified	Not specified	Canada	1997–2001	102
17	Not specified	Not specified	Scotland	1997–2001	227
Average (8–17)					**218**

Notes: As a useful yardstick one can use the suicide rate among males in the United States between 1979 and 1998 which was about 20 per 100,000. Suicide rates of inmates are highly dependent upon the time, T, they have spent in prison since their incarceration. A detailed study based on 339 suicides that occurred in the United States in 1986 found that 51 percent of the suicides occurred in the first 24 hours of incarceration. This observation is consistent with the interpretation of suicide as resulting from a severing of social ties. In the statistics published in countries other than the United States, the time of incarceration is not specified. However, since inmates incarcerated for less than one year are in greater number than those incarcerated for longer durations, one would expect the former to predominate. Therefore it is not surprising that the order of magnitude of suicide rates is more or less the same everywhere (one exception is Saxony).

Sources: 1: DCJ Report (Tables 7, 9, 11); 2: Hayes and Rowan (1988, p. 4); 3–5: Hayes and Rowan (1988, p. 52–3); 6: DCJ Report (table 1), Hayes and Rowan (1988, Table 2); 7: http://www.mces.org/Suicide_Prisons_Jails.html; 8–10: Legoyt (1881); 11: Correctional Service of Canada (1992); 12: Corrections Department of New Zealand (2003), http://www.corrections.govt.nz; 13: HMPS (2001); 14: Baron-Laforet (1999), Bourgoin (1999); 15–17: same as 12.

before the shift to confinement and to the difficulty of building new ties once in confinement. In the trauma interpretation one would argue that the shift from prison to confinement creates a shock which accounts for the higher suicide rate. At this point it is still difficult to decide which interpretation is correct, but some light will be thrown on the question by considering the so-called "silent system" that was used in the penitentiary system in the nineteenth century.

11.2.3 The silent system

Starting from the premise that prisoners learn criminal ways from each other and that isolation, on the contrary, would put them face to face with their conscience, prison reformers introduced the silent system (Villermé 1820). Architecturally, these penitentiaries were designed to minimize contact between inmates and between inmates and staff. During most of the day prisoners were kept isolated in individual cells; exercise was permitted only in small solitary yards whose high walls prevented contact with fellow inmates (Telzrow 2002). In 1848, the English writer Charles Dickens was authorized to visit the Pennsylvania penitentiary. In his account he wrote: "A prisoner is a man buried alive" (Dickens 1842). In such conditions the network perspective would lead us to expect high suicide rates.[2] This was indeed the case although the evidence available is fairly scant.

- In one of his lectures, the prominent British lawyer Louis Blom-Cooper (1987) declared that after the adoption of the silent system in Britain the suicide rate among prisoners was as high as 1,760 per 100,000.
- High rates are reported in Table 11.3 for Belgium and Saxony, two countries which were among the first to adopt the silent system.

These high rates are difficult to explain in the trauma interpretation, for in the silent system confinement is not a punishment but rather the permanent condition of prisoners.

In the next section we will see that the data available about suicide in prison may give us some insight into the time it takes for a network of social ties to form.

11.3 Effect of a rearrangement of social ties

A first set of evidence is offered by British remand centers. These are jails where prisoners are held while waiting for their trials. Usually, stays in remand

---

[2] Naturally, the actual occurrence of suicides is conditioned by the availability of means for committing suicide. If there are no sheets, towels, glasses, windows, or anything that could be used as a knife, committing suicide becomes nearly impossible. However, living in such conditions may affect the prisoner's mental health to the point that he is likely to commit suicide as soon as he is allowed to leave the confinement cell.

centers are shorter than six months. As a result, their population is in the majority composed of "new" prisoners. The following table shows that the transient state which characterizes these prisoners results in suicide rates which are about three times as high as the average prison rate.

Type of prison	Annualized suicide rate (per 100 000 of average daily population), 1996–8
Remand centers (stays shorter than 6 months)	389
Total prison population	119

Source: Marshall *et al.* (2000).

This observation is interesting in two respects. (i) The fact that the suicide rate in remand centers is higher than in solitary confinement is at first surprising. Probably the simplest explanation is that prisoners have more means at their disposal for committing suicide in remand centers than in solitary confinement cells. (ii) The analogues in the United States of the British remand centers are the jails. But whereas in remand centers stays are limited to six months, in jails they are limited to one year. This is probably the reason why suicide rates in remand centers are twice as high as in jails.

Moving from one prison to another In a study done by MacDonald and Sexton (2002) we learn that suicide rates increase *every time* a prisoner is moved from one prison to another. If time is counted from the day a prisoner arrives in a prison, the data show that there are approximately as many suicides during the first 6 days as during the following 53 days. In other words the annualized suicide rate is about nine times higher in the first 6 days than in the two following months. This observation should be considered as fairly robust because it relies on a set of data which covers the 11 years 1990–2000.

Release from prison Normally one would not interpret the fact of being released as a trauma. In the tie perspective the surge of suicides in the first week (Table 11.4) has a natural interpretation as being caused by a transient readjustment of social ties. However, Table 11.4 also reveals that there is a surge in accidental and natural deaths. The following explanations may be useful for the interpretation of these data.

- Prisoners with a fatal illness are often released in their last weeks of life.
- While the supply of alcohol and drugs is limited in prison, all of a sudden intoxicants become available. In a sense, however, their excess consumption can be seen as a kind of suicidal behavior.

Table 11.4 *Number of deaths per week of ex-prisoners in the weeks after their release*

Situation	Number of deaths per week					
	Suicide		Natural causes		Accidents	
Week 1	4.00	[1.00]	4.00	[1.00]	13.0	[1.00]
Weeks 2–4	1.70	[0.42]	2.33	[0.57]	8.3	[0.64]
Weeks 5–12	0.87	[0.22]	1.63	[0.40]	6.2	[0.47]
Weeks 13–24	0.54	[0.13]	0.66	[0.16]	5.9	[0.45]

Notes: The numbers within brackets show the data in normalized form (first week = 1). At first sight one may be tempted to interpret the decrease in the number of suicides per week as reflecting a transient state marked by a reorganization of social ties. However, the parallel observations for deaths by natural causes and by accidents clearly require different explanations. In the case of death by natural causes a key factor is that terminally ill prisoners are released so they can die at home or in hospital. Alcohol is also an important factor.
Source: Sattar (2001, p. 34).

- Life in prison provides a protection against many forms of accidental death, such as being run over by a car or drowning. Furthermore, it is more difficult to commit suicide in prison than outside.
- The ex-prisoners considered in the study on which Table 11.4 is based were under community supervision. However, the periods of supervision were limited in time and when coming to their end, the persons under consideration dropped from the statistical sample. Any death occurring afterward would not be recorded. Such a statistical procedure naturally records a higher proportion of deaths occurring shortly after release than would be the case if all deaths were recorded no matter how long after release they happen.

These explanations account for the high rate of mortality by illness or accidents in the days following the release. However, they do not invalidate our observations regarding the surge in suicides as reflecting a rearrangement of social ties.

Is there a similar effect in physics for interactions between molecules? When two liquids *A* and *B* are mixed there is a transient state during which new *A–B* bonds are established. What is the duration of this transient state? One should distinguish two characteristic times, one at molecular level and the other at macroscopic level. At the molecular level the characteristic time of bond rearrangement is very short, typically of the order of 10^{-9} seconds. For instance, when two molecules interact through the attraction due to induced densities of charge (the so-called London interaction) this interaction changes every 10^{-10} seconds; similarly it can be recalled that in a gas at room pressure and temperature the mean time between collisions is of the order of 10^{-10} seconds (Reif 1965, p. 471). However, before a new *A–B* bond can be established the molecules must be close enough. Apart from diffusion, there is no natural mechanism which ensures the mixing of two liquids, and diffusion can take a very long time (of the order of several days), especially if the densities of the liquids are fairly different. This *macroscopic time* only plays a role when one considers the mixing of two macroscopic volumes of liquid; if a few *A* molecules are introduced into the *B* liquid, they are immediately in close contact with *B* molecules. This is what happens in the remand centers considered above in so far as the prisoners arrive one by one. However, the macroscopic mixing time plays a greater role when immigrants form communities which have little contact with the population of the country where they are established.

The previous observations showing a sharp decrease in suicide rates over a time scale of a few weeks suggest that the characteristic time for the establishment of social ties in prison or in remand centers is of the order of several weeks. The sudden surges in suicide rates after an arrest or after a shift from one prison to another suggest that after the ties have been cut the peak in suicides which follows occurs within a few days or, at most, a few weeks. If this order of magnitude is correct one would expect the same characteristic time for the suicide of young widowers. In other words, the mean time between the death of the wife and the suicide of the widower should be of the order of a few weeks, a prediction which is indeed confirmed by observation (Bojanovsky 1980).

We now turn to studying the effect on suicides of the social disruption experienced by immigrants.

11.4 Effect on suicide of the social disruption experienced by immigrants

In the previous section we studied the effect of social isolation when a person is taken into custody. Naturally, this is only one of several possible mechanisms leading to a disruption of social ties; the process of emigration that we consider in this section is another. Back in the nineteenth century, when an individual or a family emigrated from a European country (say Italy for instance) to the United States it implied a sharp interruption in contacts with the relatives and friends left behind. Furthermore, until the language barrier was surmounted it was not easy to establish social ties with American people, except of course with other Italian immigrants. This was certainly an important factor in the creation of communities mainly composed of immigrants such as the "Little Italy" district in South Manhattan.

From a network perspective one would expect the suicide rate of immigrants to be inflated by the weakening of their social ties. Figures 11.3a and b show that the data points are above the line $y = x$, which means that the suicide rates of immigrants are higher than in their respective countries of origin. In addition,

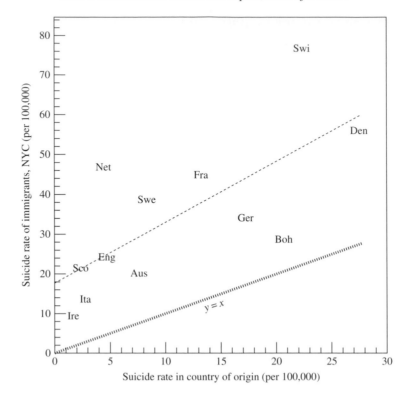

Fig. 11.3a   Suicide rates of immigrants in New York City (1870–1880). Each country abbreviation corresponds to a group of immigrants born in that country and who committed suicide in New York City ("Boh" means Bohemia, "Sco" means Scotland, the other labels are fairly clear). The suicide rates in New York City are correlated with (but higher than) the suicide rates in the countries of origin. The correlation is equal to 0.70 and the slope of the regression line is 1.5 ± 1. The line $y = x$ is shown for the purpose of comparing the rates in New York with previous rates. The high dispersion is due to the fact that some of the suicide numbers are fairly small, for instance there were only 5 suicides from Denmark, 8 from Sweden and 10 from Italy, as compared with 626 from Germany. *Sources: Suicide rates in NYC: Nagle (1882); suicide rates in countries of origin: Durkheim (1897), Krose (1906).*

the graphs show that there is strong correlation between the rates in the United States and the rates in the countries of origin; the correlation is equal to 0.70 for Fig. 11.3a and 0.85 for Fig. 11.3b. The regression coefficients are given by

$$1870–1880 \quad s(\text{NY}) = as(\text{origin}) + b \quad a = 1.5 \pm 1.0 \quad b = 18 \pm 8$$
$$1910 \quad s(\text{US}) = as(\text{origin}) + b \quad a = 1.3 \pm 0.6 \quad b = 6.9 \pm 4$$

where $s(\text{origin})$, $s(\text{NY})$, $s(\text{US})$ denote the suicide rates in the country of origin, in New York City and in the United States respectively.

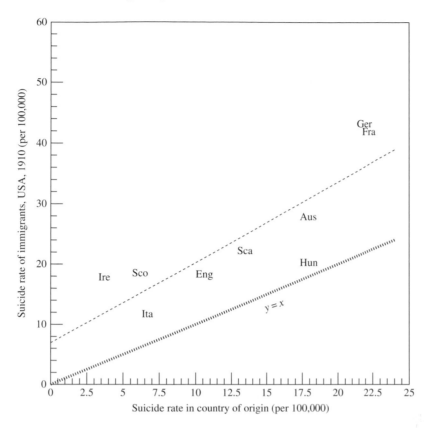

Fig. 11.3b Suicide rates of immigrants in the United States (1910). Each country abbreviation corresponds to a group of immigrants born in that country and who committed suicide in the United States ("Hun" means Hungary, "Sca" means Scandinavia, the other labels are fairly clear). The suicide rates in the United States are correlated with (but higher than) the suicide rates in the countries of origin. The correlation is equal to 0.85 and the slope of the regression line is 1.3 ± 0.6. The smallest number of suicides is 47 (Scotland) which explains why the dispersion is smaller than in the previous graph. *Sources: Suicides in the United States: Mortality Statistics (1910, pp. 586–95); foreign born population by country of birth: Historical Statistics of the United States (1975, p. 117); suicide rates in country of origin: World Health Organization (1996).*

This observation has an important implication. The suicide rates of groups of immigrants appear to be a cultural attribute just like their language, traditions or diet. We do not yet have a comprehensive explanation of this effect, but in a sense it is consistent with the fact that immigrants tend to reproduce their former social networks in the new country where they have settled. Naturally, it would be of great interest to be able to follow the changes in their suicide rates in the course of time as they become better integrated. We will come back to that question at the end of the chapter.

According to our previous argument the suicide rates of immigrants at the time of their arrival are disconnected from American suicide rates. They can be lower or higher. An example of the first kind is provided by Italians immigrants: in their country of origin their suicide rate is 6.3 per 10^5, through immigration it jumps to 11.2 which however is still lower than the rate for Americans for (15.3); another example of this kind is provided by Hispanic immigrants which we consider more closely in a moment. Examples of the second kind are provided by most other European countries listed in Fig. 11.3.

Incidentally, it can be noted that the high correlation between suicide rates in the countries of origin and in the United States proves that suicide statistics are indeed trustworthy in the sense that the statistical estimates measured in the countries of origin are corroborated by independent estimates made in America. This observation invalidates most of the reservations expressed by some authors such as Douglas (1967).[3]

During the nineteenth century and during the first half of the twentieth century most immigrants came from European countries where (with the exception of Ireland and Italy) suicide rates were fairly high. Thus, immigration was a factor which tended to raise suicide rates in the United States. In contrast, in the late twentieth century immigrants came mainly from areas such as Mexico and Central America where suicide rates are much lower than in the United States; this immigration therefore tended to lower suicide rates in the United States. In the next section we investigate this effect more closely.

11.5 Effect of immigration on suicide rates in the country of destination

As immigration is only one of the factors which affect suicide rates, it may not be easy to identify the impact of this factor; therefore, in accordance with the extreme value methodology we consider historical episodes characterized by a huge influx of immigrants.

11.5.1 Immigrants with high suicide rates

In the case of the United States, the most favorable episode is the period 1845–65. Indeed, it was in this period that immigration reached its highest rates, culminating at 18 per 1,000 in 1854 (other local maxima were 15 in 1882 and 12 in 1914).

---

[3] For a useful discussion of the arguments presented by Douglas see Besnard (1976) and Baudelot and Establet (1984).

Moreover, during 1845–65, most of the immigrants came from Germany where suicide rates were particularly high. Thus, we are in the best conditions to observe the impact of immigrants on US suicide rates.

For the nineteenth century, suicide statistics are available only in a few registration cities. Fortunately, New York City, the arrival port of most immigrants, is one of them. Figure 11.4 shows that suicide rates in New York are well correlated with immigration rates. Note that the curve of the suicides has been shifted back three years, which means that on average the suicide of new immigrants occurs three years after their arrival.[4]

Is the magnitude of the change in suicide rate compatible with the magnitude of the influx of immigrants? We do not know how many immigrants stayed in New York City after their arrival. For the sake of simplicity let us make the (provisional) assumption that all immigrants remained in New York City. During the 6 years 1849–54 the average annual immigration rate in the United

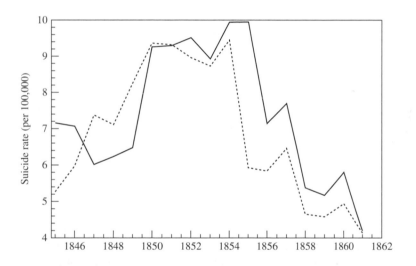

Fig. 11.4 Suicide rates in New York City compared with immigration rates in the United States. The left-hand scale applies only to suicide rates (the scale for the immigration rate is not displayed). Solid line: suicide rate in New York City (the curve has been translated back three years); broken line: immigration rate in the United States (per 1,000 of US population). The correlation is 0.67 and the regression reads: suicide $= a(\text{immigration}) + b$, $a = 0.89 \pm 0.13$, $b = 4.5 \pm 0.6$. *Sources: Suicide rates in New York City: Nagle (1882); immigration rate: Historical Statistics of the United States (1974, p. 106).*

---

[4] Three years is a time lag which is much longer than the few weeks observed in shifts from one prison to another, which suggests that the three years correspond to a macroscopic mixing time. Immigrants did not come alone and for a while their fellow countrymen provided substitutes for the links that were severed through their departure.

States was 13 per 1,000, which over the 6 years represents a cumulated number of 13×6 per $1,000 = 7.8\%$. As the population of New York City was about 8 times smaller than the US population the rate of immigration was 8 times greater, that is $7.8 \times 8 = 62\%$. With these simplifying assumptions a sample of 100,000 New York residents in 1849 would comprise 162,000 people in 1854 of whom 62,000 would be immigrants, that is to say a proportion of $62/162 = 37\%$.[5] For the sake of simplicity we also assume that all immigrants have the suicide rate of the Germans (who composed the largest group), that is to say about 35 per 100,000.

Starting from a suicide rate of $s_{1849} = 6$ we get in 1854 a suicide rate of $s_{1854} = 0.63 \times 6 + 0.37 \times 35 = 16.7$. If we assume that only one-half of the immigrants remained in New York City during the 6 years, we get instead $s_{1854} = 0.82 \times 6 + 0.18 \times 35 = 11.2$, which is close to the rate actually observed in 1854. In short, the orders of magnitude seem to be plausible. Furthermore, if we assume that the rate at which immigrants leave New York City is fairly constant the parallelism between the two curves after 1855 gets a natural interpretation.

One may wonder what effect emigration has on suicide rates in the country of origin. A simple argument shows that it is likely to push suicide rates up. In a general way, suicide rates are known to increase with age and this is especially true for males. As emigrants are mainly young people under 40 the country which they leave will experience an increase in the proportion of its population which is over 40. This will result in a higher crude (as opposed to age-standardized) suicide rate.

We now turn to Hispanic immigration where immigrants have a suicide rate which is lower than the average suicide rate in the United States.

11.5.2 Effect of Hispanic immigration on US suicide rates

Between 1990 and 2002, the US suicide rate fell from 12.4 to 11.0 per 100,000. Some researchers attributed this decline to the growing use of anti-depressants and particularly of Prozac.[6] In recent years there has been a growing tendency to explain suicide by psychological factors. In the present section we propose a testable explanation of the fall in suicide rates as being due to Hispanic immigration. First, we give some crucial parameters about Hispanic immigrants and in a second step we present a statistical test based on suicide rates in Californian counties.

---

[5] In 1875, the population of New York City comprised 44% foreigners.
[6] See the following articles: "Suicide rate down in the era of Prozac" (*The Nation*, February 5, 2005); "Antidepressants save lives" (http://washingtontimes.com).

The suicide rate of Hispanics in the United States was 5.6 in 2001 and 5.0 in 2002.[7] For the purpose of comparison the suicide rate in Mexico was about 3.0 per 100,000 (Puentes-Rosas *et al.* 2004). It is of interest to observe that the ratio $5.3/3.0 = 1.8$ is consistent with the data about immigrants from other countries given in Figs. 11.3a,b.[8]

As the rate of 5.3 is only one-half of the US rate, an influx of Hispanics lowers the US rate as surely as the addition of cold water to hot water lowers the temperature of the mixture. Furthermore, the fall in suicide rates should be greater in areas where there has been a large influx of Hispanics. Figure 11.5a provides a test based on Californian counties. For each of the 58 counties the plot shows (i) the percentage of Hispanics in its population in 1996; (ii) the average suicide rate over the five-year interval 1994–8. The coefficient of correlation is -0.76, which confirms the connection between the level of suicide rates and the presence of Hispanics.

A more detailed examination shows that in fact the relationship is not linear. The dotted curve represents an average performed over successive groups of five points. How can that non-linearity be interpreted? Let us denote by t_1 and t_2 the suicide rates of non-Hispanics and of Hispanics respectively. If these rates were independent of the proportion of Hispanics (x) the total suicide rate (t) would be given by

$$t = (1 - x)t_1 + xt_2 = (t_2 - t_1)x + t_1$$

Thus, the fact that the relationship is *not* a straight line shows that t_1 and t_2 are in fact functions of x. For the sake of simplicity let us assume that these functions are linear:

$$t_1 = \alpha_1 x + \beta_1 \qquad t_2 = \alpha_2 x + \beta_2$$

As a result, the function $t = t(x)$ becomes the equation of a parabola:

$$t = x^2(\alpha_2 - \alpha_1) + x(\beta_2 - \beta_1 + \alpha_1) + \beta_1 \qquad 0 \le x \le 1$$

Because of the concavity of the curve in Fig. 11.5a one obtains the condition $\alpha_2 > \alpha_1$, a prediction which it will be possible to test when suicide rates become available separately for non-Hispanics and Hispanics at county level.

As a second test of the relationship between the fall in suicide rates in California and the influence of Hispanics we propose a longitudinal analysis for the two counties of Los Angeles and San Diego. Figure 11.5b shows the total suicide rate

---

[7] *Morbidity and Mortality Weekly Report* (MMWR), Centers for Disease Control, June 11, 2004, p. 478–481; *National Vital Statistics Reports* **59**, 5, Oct. 12, 2004).

[8] Although not all Hispanic immigrants came from Mexico, immigrants from Mexico represented the largest national group; moreover the suicide rates in other Latin American countries were of the same magnitude as in Mexico.

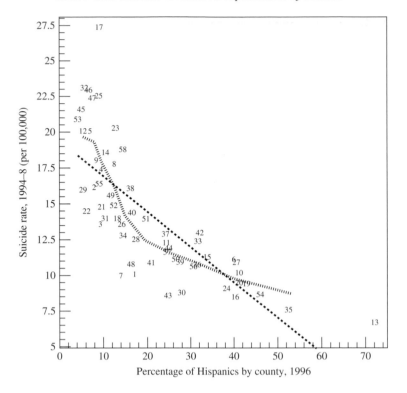

Fig. 11.5a Relationship between percentage of Hispanics and suicide rates at county level. Each number refers to one of the 58 counties of California listed in alphabetical order. The correlation is -0.76 and the linear regression reads: $y = (-0.25)x + 19$. The dotted curve gives an average over successive groups of 5 points. *Sources: Percentage of Hispanics: USA Counties 1998 (website of the US Census Bureau). Suicide rates: WONDER database (website of the Centers for Disease Control).*

as a function of the proportion of Hispanics for the four years 1980, 1990, 1996 and 2000. The points for 1980, 1990 and 1996 fall fairly well on the regression line whereas the points for 2000 are markedly below the regression line. This may be due to the fact that the data for 2000 underestimate the number of Hispanics. It is usually assumed that in addition to the 30 million legal Hispanic immigrants there are about 11 million illegal immigrants (*New York Times*, May 2, 2006). Adding 30% to the number of Hispanics of course changes the percentage of Hispanics (i.e. the *x*-variable) but it also has an effect on the suicide rate (i.e. the *y*-variable). The main point is that the number of suicides is not likely to be changed, because death certificates are established for any death whether of a legal or an illegal immigrant. However, the denominator of the suicide rate refers to the total population which is of course the registered population. Therefore

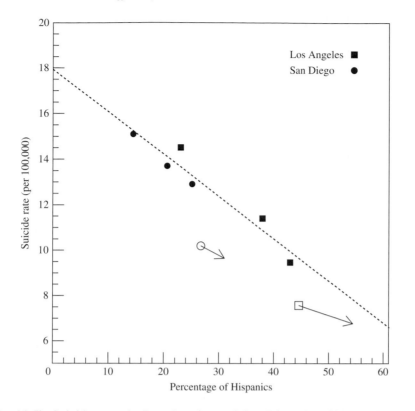

Fig. 11.5b Suicide rates in Los Angeles and San Diego in 1980, 1990, 1996, 2000. The squares and circles refer to Los Angeles and San Diego respectively; if one excepts the open symbols (which concern 2000) the correlation is −0.96. A possible reason which may explain why the data for 2000 are outliers may be an underestimation of the numbers of Hispanics. The arrows indicate likely corrections as explained in the text. *Sources: Same as for Fig. 11.5a.*

the correction will result in reduced suicide rates. The expected corrections are indicated by arrows in Fig. 11.5b.

11.6 Non-linear mixing relationships

In this section we are interested in the suicide rate of a population which is composed of two separate subgroups. From a network perspective, this question can be stated in the following general form. If two separate networks A and B are characterized by intensive parameters p_A and p_B, what happens when these networks are allowed to interact; in particular how is the new parameter $p_{A \cup B}$ of the global system defined in terms of p_A, p_B? A related question is how long it takes for the mixing of A and B to result in an equilibrium situation for the new network $A \cup B$. The first question is an equilibrium problem while the second is a time-dependent issue.

To get some insight into this problem we need a dataset which fulfills the following conditions. (i) The two subgroups A and B must have suicide rates which are sufficiently different. (ii) The distinction between A and B should not be based on religion for one knows that, at least after the nineteenth century, the religion of a group is little affected by the process of integration. There are very few countries which publish suicide rates for different components of their populations. It is thanks to the detailed statistical data published by the United States for the Black and White components of the American population that we will be able to study the suicide rates in these two subgroups. We begin our discussion with a cross-sectional analysis summarized in Figs. 11.6a and b.

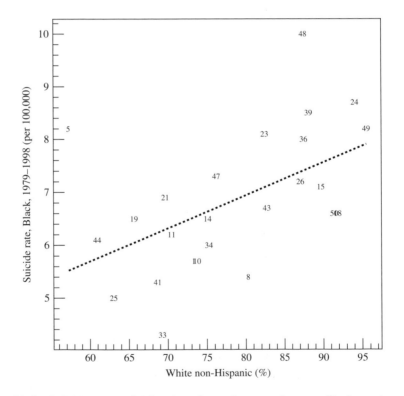

Fig. 11.6a Suicide rates of Afro-Americans in several states. Horizontal scale: percentage of White non-Hispanic people in 25 states; vertical scale: average suicide rate of Afro-Americans over the 20 year period 1979–98. The correlation is 0.52 (confidence interval at 95% probability level is (0.15, 0.75)); the slope of the regression line is 0.062 ± 0.04. *Sources: Same as for Fig. 11.5a.*

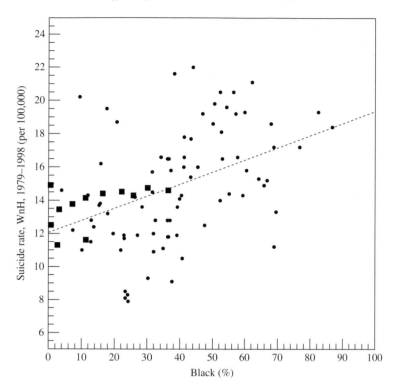

Fig. 11.6b Suicide rates of White non-Hispanics in American states and in the counties of Mississippi. Horizontal scale: percentage of Afro-Americans; vertical scale: average suicide rate of White non-Hispanics over the 20-year period 1979–98. Each square represents the data for a state whereas the dots correspond to the counties of Mississippi. The counties of Mississippi were included because this state has the highest proportion of Afro-Americans. Only states with less than 2% Hispanics were included in the sample so that the suicide rate of Whites given by the data can be identified with the suicide rate of White non-Hispanics. The correlation is 0.44 (confidence interval at 95% probability level is (0.26, 0.59)); the slope of the regression line is 0.073 ± 0.03. The fact that the regression is not a horizontal line suggests that in their behavior with respect to suicide individuals are sensitive to their social environment. *Sources: Same as for Fig. 11.5a.*

Figure 11.6a shows the suicide rates of Afro-Americans in 25 states as a function of the percentage of the White non-Hispanic population. These states were selected on the basis of the two following criteria.

- In order to mitigate the influence of population densities we selected states whose densities are between 20 and 200 people per square kilometer.
- In order to minimize statistical fluctuations we eliminated the states having less than 20 suicides over the period 1979–98 (which means less than one suicide per year).

How should we interpret the positive slope of the regression line in Fig. 11.6a? First we can note that any slope, whether positive or negative, demonstrates the existence of a collective effect of the social environment on the Black population. A fairly natural explanation would be to argue that Black people who live in a state with a high proportion of White non-Hispanics may adopt the ways and behavior of the people around them including their higher suicide rate. This explanation will be referred to as an immersion-contagion effect; it is a fairly anthropomorphic explanation in so far as the notions of behavior, of contagion or imitation have no clear translation in network concepts. On the basis of this argument one would expect the suicide rate of White non-Hispanics to decrease as the proportion of the Black population becomes greater. Yet Fig. 11.6b shows that it is the opposite which is observed. In short, the immersion-contagion effect does not work. To nevertheless "save" this explanation one could argue that Whites are less likely to adopt the ways and behavior of their Black neighbors. Yet even if one accepts this argument one still has to explain why the slope of the regression line is positive instead of being simply close to zero.

As the anthropomorphic explanation failed, let us try the network approach that we have been using throughout this book. A first step is to convince ourselves that what we observe in Figs. 11.6a, b is a fairly standard case in the physics of liquid mixtures. If one wishes to draw a parallel with the mixing of two liquids, the analogue of the suicide rate of population A would be the ratio (Number of molecules of A in the vapor above the liquid)/(Number of molecules of A in the liquid) $\sim$ (Partial pressure p_A of A in the vapor above the liquid)/(Concentration x_A of A in the liquid).[9] In the case of liquids which obey Raoult's law this curve would be a horizontal line. In a solution with a positive deviation from Raoult's law the curve of p_A/x_A would be an increasing function of x_B and similarly p_B/x_B would be an increasing function of x_A. In short, this case corresponds to what we observe in Figs. 11.6a, b. Thus, in order (tentatively) to get an interpretation of these graphs we must recall the meaning (in terms of interactions) of a positive deviation from Raoult's law. A positive deviation implies the following relationship between strengths of attraction (SA): $(1/2)[\text{SA in pure } A + \text{SA in pure } B] > \text{SA between } A$ and B molecules in the mixture.

We complement the previous cross-sectional studies by a longitudinal analysis of the White and Black components.

---

[9] The population A is represented by the A molecules in the liquid, the people of A who commit suicide are represented by the A molecules in the vapor above the liquid.

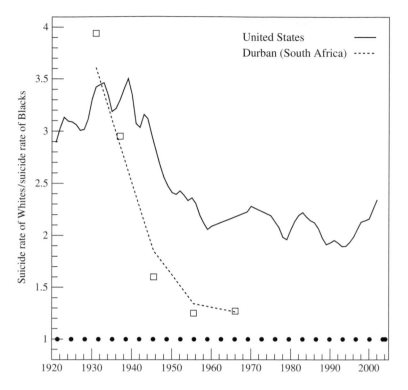

Fig. 11.7 Ratio of the suicide rate of Whites to the suicide rates of Blacks. Under the assumption of a convergence of White and Black social behavior one would expect a convergence of their suicide rates as well as of other social characteristics such as infant mortality. Observation shows that this did not happen; in spite of the fact that religious factors do not play an important role (as would be the case for Irish or Italian immigrants), the two suicide rates do not display any convergence trend over the past decades; for infant mortality (which is not shown here) there was a slow convergence until 1950 after which the two rates began to diverge. In this graph, "Whites" includes Hispanics, but until 1990 the percentage of Hispanics was small, which means that the suicide rate of Whites can be identified with the rate of White non-Hispanics. *Sources: United States, 1921–1940: Linder and Grove (1947, pp. 272–3); United States, 1940–60: Grove and Hetzel (1968, p. 373); United States, 1960–79: Statistical Abstract of the United States 1969, 1981; United States, 1979–2002: WONDER database (website of the Centers for Disease Control); South Africa: Meer (1976, p. 236); Infant mortality: Roehner (2004, p. 244).*

Longitudinal analysis Figure 11.7 shows how the ratio of White to Black suicide rates changed in the course of time between 1920 and 2002. In this time interval there has been a slow increase in the percentage of the Black population: it was 9.40% in 1920, 10.8% in 1970 and 12.8% in 2002. These slight changes certainly cannot account for the ups and downs displayed by the curve. Actually, the main message of Fig. 11.7 is that the ratio does *not* converge toward 1. As already noted this persistent gap can hardly be explained by religious factors.[10] The curve for Blacks and Whites in Durban (South Africa) shows greater convergence of suicide rates. Clearly, additional data for other countries would be welcome but, as already noted, only a few countries record suicide rates separately for different components of the population.

---

[10] In 2000 the number of Muslim Afro-Americans was estimated at 3 million, that is to say about 10% of the Afro-American population.

12

Apoptosis

In any society, whether human or non-human, there are links and interactions between individual units. A possible method for investigating these links is to consider what happens when they are severed. For instance, what will happen to an ant which is removed from its nest and kept in isolation in a place which offers the same conditions as the nest in terms of food, temperature, humidity and other factors? Will it have a shorter or longer life than the ants which remain in the nest? If, as may be expected, its life is shortened, one would like to know *how much* it is shortened. Before answering this question (which we do later on in this chapter) it is important to realize its implications. This point can be illustrated through the following gedankenexperiment. Suppose you are slicing bread and the knife slips. Your finger is cut and starts to bleed. You put on a plaster and think nothing more of it. However, some skin cells have been displaced down into the muscle tissue; if they survive and divide they will produce skin cells in a location where no skin cells should exist. Fortunately, these displaced skin cells undergo self-destruction, a mechanism known as apoptosis.[1] Other expressions such as programmed cell death (PCD) or cell suicide (Raff 1998) are also used to designate the mechanism of self-destruction.

In order to establish a connection between our previous question about ants kept in isolation and the phenomenon of apoptosis one may wonder what happens when a skin cell is removed and kept in a test tube. Martin Raff, a pioneer in the study of apoptosis, has performed numerous experiments of this kind. He was able to show that whereas cells can survive for weeks in cultures at high cell density, they undergo apoptosis at low cell density. This led Raff to the conclusion that the only thing that stops living cells from ending their lives is the constant receipt of a biochemical signal saying "Stay alive, stay alive, ..." When living cells are put

---

[1] The term "apoptosis" comes from a combination of two Greek words *apo* which means "from" and *ptosis* which means "falling"; in Greek the term apoptosis refers for instance to leaves falling from a tree.

to grow in low density culture dishes they activate their death program probably because the "stay alive" signals that they receive are too weak (Ishizaki *et al.* 1994, Raff 1998).

In this chapter we first explain in what sense apoptosis is fundamentally different from necrosis, which is another form of cell death. Then we examine why apoptosis is an important function for all organisms. A question of central importance is whether some form of apoptosis exists in populations which are not organisms. The very existence of organisms such as *Dictyostelium discoideum* (to which we come back later on) shows that the boundary between populations and organisms is not clear-cut but rather fuzzy and porous; therefore one would not be surprised to find apoptosis mechanisms also in populations. In a more general way one may conjecture that mechanisms similar to apoptosis may exist in all evolving networks for the purpose of eliminating links or nodes which are no longer serviceable.

12.1 Apoptosis versus necrosis

Cells can die in two different ways. (i) In necrosis the cell swells without any rearrangement of its internal components; eventually, the membrane of the cell is ruptured; the nucleus, DNA, mitochondria and all other internal building blocks are released and destroyed. Locally, the process results in an inflammatory reaction. (ii) In apoptosis, the cell first shrinks, its internal components are "packaged" in an orderly way before being released, engulfed and "swallowed" by neighboring cells. In short, apoptosis is an elaborate program of self-destruction which results in orderly dismantling and reuse of cell components. Necrosis and apoptosis can be distinguished without ambiguity through microscopic observation (Fig. 12.1). How long does it take for the process of apoptosis described in Fig. 12.1 to be completed? The answer depends on the type of cell but an order of magnitude is 10 to 20 hours.

Recognition of programmed cellular death as a development mechanism dates back well over 100 years (Clarke and Clarke 1996). However, it is only in the last decades of the twentieth century that significant progress has been made in its understanding. Much of our current knowledge comes from work on *Caenorhabditis elegans*, a small worm about one millimeter in length which has 959 cells. In the course of its development 1,090 cells are produced of which 131 die by apoptosis. The 2002 Nobel Prize in Physiology or Medicine was awarded to Sydney Bremmer, H. Robert Horwitz and John Sulston for their discovery concerning genetic regulations of programmed cell death.

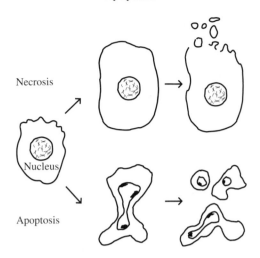

Fig. 12.1 Necrosis versus apoptosis. Necrosis is an explosive cell death that results from severe injury and is characterized by swelling, cell rupture and release of the cell components in a way which can damage nearby healthy cells. Note that the DNA remains dispersed throughout the nucleus. Apoptosis (also called cell suicide) is a controlled death process in which the cell components are "packaged" and ingested by nearby cells.

12.2 Role of apoptosis in the development of multicellular organisms

Apoptosis fulfills a number of important functions, but before examining them more closely it may be of interest to present an instance in which apoptosis seems almost purposeless. We already mentioned that because of its simplicity *C. elegans* provides an ideal testing ground for the mechanisms leading to apoptosis. It was found that two specific genes control whether or not a cell should survive. When these genes were inactivated, the 131 cells which normally undergo apoptosis were able to survive in an adult worm. It turns out that having 1,090 cells did not affect its survival capability. Thus, in this specific case, we see that apoptosis has no clearly defined function. However, there are many other cases in which apoptosis fulfills essential functions. One can mention the following.

- The resorption of the tadpole tail at the time of its metamorphosis into a frog occurs by apoptosis. Similarly, the formation of the fingers and toes of the fetus requires the removal by apoptosis of the tissue between them.
- One method by which lymphocytes eliminate virus-infected cells is by inducing apoptosis.
- Cells which for some reason are moved away from their normal location disappear by apoptosis.
- Cells which are damaged beyond possibility of being repaired and restored are eliminated by apoptosis.

It has been estimated that in a human organism from 50 to 70 billion cells die each day by apoptosis. It is interesting to compute the corresponding annual death rate. A human body has approximately 100,000 billion cells which leads to a rate of $365 \times 60 \times 10^9/(10^{14}/10^5) = 21,900$ per 100,000. This rate is about 1,000 times larger than the average annual suicide rate in human societies. However, we will see that suicide is only one of several mechanisms by which the functions normally performed by apoptosis can be achieved in human societies.

12.3 Apoptosis in plants

For fruit trees there is a kind of apoptosis mechanism through which fruits which are damaged are eliminated at an early maturation stage. As an example let us consider the premature drop of citrus. At first sight one may attribute this phenomenon to weather conditions, for instance rainfall or strong winds. However, a closer examination shows that weather is not the real cause. A first indication is the fact that premature fall does not necessarily affect fruits growing on high branches which are the most exposed to wind gusts. Moreover, the fallen fruit reveal a split at the navel end (that is to say the end of the citrus opposite to the stem end). Upon cutting the citrus one sees that there is a black discoloration due to a fungal organism (Grafton-Cardwell 2005) which suggests that it is the invasion of the fruit by a pathogen which provoked the separation of the fruit from the tree in the same way as virus-infected cells are eliminated by lymphocytes. Other cases are reported in the literature which confirm the idea that the fruits which fall from trees are not selected randomly. For instance, it has been observed that prematurely fallen apples on average contain a smaller number of seeds than the apples which remain on the tree (Ulrich 1944).

12.4 Apoptosis in populations

Does apoptosis also occur in populations of social insects, in herds of mammals (e.g. herds of antelopes) or in human societies? The question is more difficult to answer than for cells because one can no longer rely on the clear-cut observational distinction between necrosis and apoptosis. There are two ways to get around this difficulty. The first is to observe entities such as social amoebae which are intermediate between populations and organisms. The second is to use Martin Raff's isolation technique.

- Amoebae are small living organisms consisting of only one cell. In ordinary conditions they behave as individual and fairly independent entities. However, under specific circumstances, for instance when subject to food deprivation, they may aggregate and form organisms of which they constitute the cells. In the case of the amoeba

Dictyostelium discoideum one observes the formation of a slug-like organism in which the amoebae undergo a process of differentiation into spore and stalk cells. The interesting point from our perspective is the fact that the stalk is composed of dead cells that have undergone programmed death much in the same way as in cellular apoptosis. The fact that apoptosis does not seem to occur in populations of more or less independent amoebae but appears as soon as a global entity is formed clearly shows that apoptosis capability is a property of global organisms. In a sense it can be seen as the signature of a high level of interdependence between the units composing the organism. Below we test this conjecture in the case of populations of social insects.

- It may be remembered that in Raff's experiments, when cells were grown in low density cultures the cells activated their apoptosis program and died. In May–July 1944, Pierre-P. Grassé and Rémy Chauvin performed similar experiments with insects. Their outcomes are summarized in Fig. 12.2. To the best of our knowledge, this kind of experiment has not been repeated, which is fairly surprising given the great interest in these questions in recent decades.

Grassé and Chauvin describe their experiments as follows. Bees from the same bee hive were captured on leaving the hive and were put into small boxes (150 cm^3 in volume) fitted with a wire mesh lid. Different boxes were prepared with a single bee or with 2, 3, 5 or 10 bees. All of them received plenty of food in the form of water sweetened with saccharose. The authors say that they used a total of 250 bees but they do not specify how many boxes of each sort were prepared; therefore we do not know the margin of error of the experimental points. A similar experiment was performed with ants. In this case a total of 105 worker ants from three different nests of a species of *Leptothorax* were used. The boxes consisted of small cavities in a block of plaster whose humidity was maintained. The ants were nourished with wood powder, saccharose and tiny bits of grasshoppers. A second experiment was performed with ants of the species *Formica rufa*. A total of 120 ants were captured from the same nest and kept in groups of 1, 2, 3, 5 or 10. Every two days they received fragments of a ripe apple. Similar experiments were performed with wasps (*Polistes gallicus*) and with termites (*Reticulitermes lucifugus*).

The life expectancy of the insects in the isolated subgroups of 1, 2 and 10 is described in Fig. 12.2. Two observations can be made.

(1) In all cases the life expectancy of a single individual does not exceed 20 days.
(2) Except for the wasps, the life expectancy is greater for the group of 10 than for the groups of 1 and 2 individuals. This suggests that, as in Raff's experiments, the life expectancy is density dependent. As a conjecture we suggest that it is in fact a function of the strength of the interaction. When the insects are in their nest the interaction is maximum, it subsists in weak form in groups of 10 and it is minimum for single individuals.

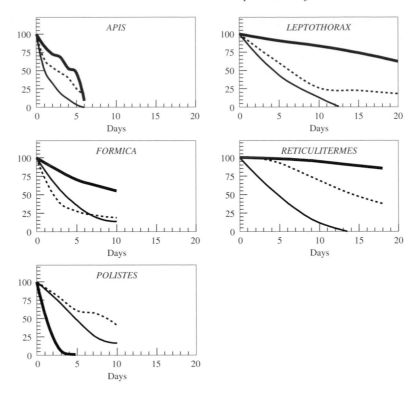

Fig. 12.2 Effect of removing insects from their colony. Several groups of 1, 2 or 10 insects were removed from their nest and kept in isolation. They received plenty of food and were kept in conditions which matched as closely as possible those in the nest. The vertical scale shows the percentage of surviving insects in each group. The thin solid lines represent the single insects, the broken lines represent the groups of 2 insects and the thick solid lines represent the groups of 10 insects. Initiated in 1944 by Grassé and Chauvin this kind of experiment may provide a methodology for gauging the strength of the interactions in the colony. *Apis* designates a species of bees, *Leptothorax* and *Formica* are two species of ants, *Reticulitermes* are termites and *Polistes* are wasps. The insects in isolation received plenty of food and were kept in conditions which matched as closely as possible those in the nest. *Source: Grassé and Chauvin (1944).*

If the previous conjecture is correct, this kind of experiment would give us a method for measuring the strength of the interactions in a population. Thanks to the studies of entomologists and ethologists in the past 50 years we have gained a good understanding of the means of communication of social insects. Yet we still know very little about the strength of the interactions. There are certainly marked differences in interaction strength between different species of ants but we are unable to estimate them. For instance, we know that the exchange of food between individuals, an interaction that entomologists call trophallaxis, is an important way of communication. We also know that trophallaxis is more important for termites and ants than for bees. Yet the results in Fig. 12.2 show that, of the five species, bees are the most vulnerable to isolation. This would suggest that apart from trophallaxis bees have other ways of communication which are as important.[2] Another intriguing question concerns the results obtained for the wasps. Grassé and Chauvin say that they repeated this experiment four times and that the results consistently showed that a larger subset is *not* conducive to a longer life expectancy. How can one understand this exception? Grassé and Chauvin left the question open until additional experiments provide some clues. For instance, it would be of interest to know if these results hold for other species of wasps or if they are specific to *Polistes gallicus*.

Although the pioneering work of Grassé and Chauvin is cited in some recent research papers, it is not seen as opening an avenue of research that should be pursued. This attitude is probably due to the fact that there has been a notable shift in the interest of researchers. Nowadays, the emphasis is on detailed descriptions of interaction mechanisms[3] rather than on the kind of comparative studies initiated by Grassé and Chauvin. From a system science perspective this methodological choice appears perplexing. Are detailed descriptions really conducive to a better understanding? History provides myriads of detailed descriptions from which little overall understanding can be gained unless the information is analyzed from a comparative perspective. In sociology, the fact that we are able to know social interactions in all their minute details proves more an embarrassment than an asset. In all these fields choosing the "right" level of description is a crucial step. Physicists and chemists restricted themselves to a fairly schematic view of the complex mechanisms of interaction at the molecular and atomic level; in contrast, all their efforts were devoted to developing a comparative analysis of physical phenomena which highlighted the central role of a few simple mechanisms. As one knows, this approach proved highly successful.

12.5 Apoptosis in groups of mammals

What is the effect of social isolation on life expectancy in groups of mammals? As a preliminary observation it can be noted that many mammals such as gazelles or antelopes live in fairly small groups of 10 to 15 individuals and form large herds only in specific circumstances, for instance during their migrations. The same observation applies to birds. The large groups of birds that can be seen in nesting areas are gatherings of individuals rather than societies of highly interconnected entities. The colonies of prairie dogs in the United States or Mexico are one of

---

[2] We know that bees have elaborate communication techniques for sharing information about floral patch feeding sites, but it is not clear if these ways of communication are also used for other purposes.

[3] The papers by Boulay *et al.* (1998, 1999, 2000) are representative of this approach.

the few groups of mammals which compare with colonies of social insects. Yet even large "towns" of prairie dogs contain less than 5,000 individuals whereas nests of ants or termites may contain several million individuals. Moreover, they are an assemblage of interconnected family units rather than wholly integrated societies.

There are reports of high death rates for dolphins kept in marinas, especially during the first months of captivity. However, it is difficult to know whether these deaths are due to social isolation or rather to the small size of the pools. One should remember in that respect that dolphins swim up to 80 kilometers a day and dive to depths of 200 meters.[4]

12.6 Apoptosis in human societies

The main message of the previous sections is that the more integrated is an organism the more important it is to eliminate "anomalies", a word which should be taken in a broad sense and includes for instance (i) a cell which has been displaced to a wrong place, (ii) a cell which has been damaged beyond repair, (iii) a cell which has been penetrated by a virus or an ant which has been infected by a parasite. We underlined Raff's interpretation in which "stay alive" signals are received permanently by any cell. The experiments performed by Grassé and Chauvin suggest that there are similar mechanisms in populations of social insects. In this last section we wish to discuss briefly the possible implications for human societies.

It is not unreasonable to suppose that the existence of highly integrated organizations such as churches, armies or firms would be imperiled by the presence and growth of "anomalies", but what precise meaning should we give to this word? In the context of a church, a person who does not share the common creed certainly represents an "anomaly" which should be removed. The removal can occur in different ways.

- Excommunication is the most common way. For instance, in Amish communities the concept of church discipline plays a key role.[5] When unrepentant dissidents are excommunicated they are shunned and rejected by their community even in ordinary life. Some Amish have even called for the suspension of the marital tie when one of the partners has been excommunicated; in their view, church discipline should have precedence over marriage. Confronted with such broad ostracism, dissidents have no other choice than to leave the community. In this case apoptosis takes the form of exclusion (Nolt 1992).

---

[4] As a matter of comparison, dolphins can be kept in tanks with an area of 8 meters × 8 meters and a depth of only 2 meters, under the requirements of the US Animal Welfare Act.

[5] The Amish are Anabaptists, a radical wing of the Protestant Reformation movement; their main implantation is in Pennsylvania.

- Back in the sixteenth, seventeenth and eighteenth centuries, when religion played a key role in Western societies, the removal of dissidents often led to their execution. Apart from the case of the Papal inquisition, which is well known, one can also mention the Penal Laws in England. Enacted between 1559 and 1580 they provided the penalty of high treason for the third offense which means that unrepentant dissidents could be dispossessed of all lands and goods before being executed. Moreover, from 1563 to 1736 witchcraft was an offense punishable by death in England; as an example of this kind of episode one can mention the witch hunt that took place in East Anglia between 1643 and 1647 and resulted in the execution of about one hundred witches (Gaskill 2005). For the whole of Europe the total number of witch trials known to have ended in executions is around 12,000 (Wikipedia, article "Witch trial").

- As a striking illustration of the religious climate in the middle of the sixteenth century, one can mention that a chorale in a cantata by J. S. Bach (BWV 126: *Erhalt uns, Gott, in deinem Wort*, i.e. "Lord, keep us in your word") contains the following prayer (my translation): "Lord, keep us in your word and direct death at the Pope and Turks who try to tear from his throne Christ Jesus your son."[6]

- Military discipline is fairly harsh because "dissidents" must be removed before they are able to contaminate the rest of the unit. The fact that in time of war the death penalty has been commonly used should not come as a surprise. Clearly, in such a situation exclusion or imprisonment would not be an effective deterrent.

- In contrast, firms and corporations can rid themselves of "dissidents" simply by discharging them.

- The issue is much more difficult for schools. One of the main purposes of schools is to give a degree of cultural homogeneity. Of course, unrepentant "dissidents" can be expelled from their schools but they cannot be completely excluded from the education system, at least not until they are old enough to work. This implies that the education system must develop institutions that are able to handle "dissidents". The less homogeneous the audience, the more frequent the manifestations of dissidence and the more important the institutions which are in charge of "dissidents".[7]

In short, for institutions the phenomenon of dropout, whether voluntary or forced, plays the same role as apoptosis in multicellular organisms. Can one apply the same kind of argument to societies? An illustration based on a concrete historical episode may give us an insight. Consider communities of Christians in Japan around 1600. Historical accounts show that Christians were not well accepted by the rest of the population. In such a situation there were several possible outcomes depending on the specific situation.

---

[6] The text, which is due to Martin Luther, reads in German as follows: *Erhalt uns, Herr, bei deinem Word, und steuer des Papsts und Türken Mord, die Jesum Christum, deinen Sohn, stürzen wollen von seinem Thron.* Bach's cantata was played for the first time on February 4, 1725.

[7] As an illustration one can mention the fact that for 15–17-year-old youths the dropout rate is 16% for recently arrived immigrants and 5% for immigrants who arrived in the United States during early childhood (Fry 2005).

(1) In the face of growing hostility, Dutch or Portuguese traders or settlers may have decided to leave Japan and return to Europe.

(2) Japanese Christians may have been tempted to move to isolated places and form closed communities which would have had minimal contact with non-Christian Japanese.

(3) In seventeenth-century Japan the religion of the people was to a large extent determined by the religion embraced by their lord. As a result, the defeat and overthrow of a Christian warlord had the likely effect that his subjects would return to Shintoism.

(4) In contrast to what happened in Europe for Puritans, Quakers, Anabaptists and many other religious minorities, emigration does not seem to have been a real option for Christian Japanese. Thus, if none of the previous solutions was adopted, there remained the possibility of committing suicide.

(5) Finally, persecutions and pogroms may have occurred to the effect of eliminating the Christian dissidents. History tells us that there were several episodes of this kind in the early seventeenth century, for instance in 1614, 1626 and 1637 (*Quid* 1997, p. 1277).

The previous discussion shows that suicide is only one of the possible responses to a situation of social isolation. Earlier in this chapter we noted that in a human organism the apoptosis rate is about 20,000 per 100,000. A social parallel of this rate would include all mechanisms through which "anomalies" can be eliminated; for instance in the case of seventeenth-century Japan one would have to sum up the rates corresponding to the five previous mechanisms.

13

Perspectives

In the previous chapters we pointed out that there are basically two ways to win the battle against noise. The first and by far the most effective solution is to find situations where the effect that one wishes to investigate is sufficiently strong to give a high signal to noise ratio. For instance, the effect on suicide of a male–female imbalance will be stronger in a population whose sex ratio is 20 than in one whose sex ratio is 1.2. This is what we called the extreme value technique and it played a considerable role in previous chapters. That approach is particularly useful in the first stage of a study when one wishes to identify an effect and to get a rough idea of its magnitude. In a second stage the objective is to perform the most accurate measurements of the phenomenon under consideration. In this case one resorts to the second method of reducing the influence of the background noise, namely increasing the number N of events; this means that one will collect and analyze the largest possible number of (good quality) data.

To reach these goals it is essential to have access to a great variety of large databases. Needless to say, the Internet marks a revolution in this respect. Let us illustrate this statement by two examples drawn from previous chapters.

- In Part III we used repeatedly the annual issues of *Mortality Statistics* (whose title changed to *Vital Statistics of the United States* in 1938). Each annual issue comprises quite a few volumes and totals several thousand pages. According to the electronic catalog of the network of French university libraries a fairly complete collection of this publication is available only at the national library (*Bibliothèque Nationale de France*); the availability of this periodical is probably as limited in other European countries. Moreover, the French collection is not without gaps, especially during the period of World War II. Through the Internet it is now possible to access each volume with a click; the download time for a volume of 700 pages is about three minutes and somewhat less during weekends.

- In fact, the Internet not only speeds up the research of data, it gives access to information that would have been completely out of reach otherwise. For instance, the question of the Allied occupation of Japan necessitated researches in the national archives of

241

Australia, New Zealand and the United States. Ten years ago, that would have required stays in Canberra, Wellington and Washington. Thanks to the Internet it is now possible to search online catalogs,[1] to contact archivists and get advice from them, and finally to order photocopies of the files in which one is interested. Moreover, many national libraries and archives have set up multiannual programs with the objective of making millions of books and files available online.[2]

These steps represent an immense improvement in terms of data accessibility. Naturally, one must also keep in mind that in all countries there are some sensitive data which will not be made available to researchers because of national security requirements. As a matter of fact, over the last decade these requirements have been tightened rather than relaxed; see in this respect the articles by Scott Shane in the *New York Times* of March 3 and April 18, 2006. In spite of these reservations great progress can be expected in the next decades.

Throughout this book we tried to convince the reader that the segmentation which exists at present in the social sciences is a great obstacle to sound and systematic studies. For instance, the phenomenon of dissemination of behavior patterns is basically a *sociological* phenomenon but it has great implications for *economics*; moreover, if one wishes to compare a large number of instances one has by necessity to include all *historical* cases for which sufficient information is available. Because econophysicists[3] are able to cross academic borders much more easily than economists, sociologists or historians, they are in a favorable position to carry out such a research agenda. Never before has the situation been more favorable. Let us express the hope that we will be able to seize this opportunity.

---

[1] At the time of writing (July 2006) only about 10% of the holdings are listed in online catalogs but this proportion will increase in coming years as cataloging progresses.

[2] For instance, more than 6 million books will be made available online over the next five years in the European Union.

[3] Throughout this book we used the term "econophysics" in a broad sense which, apart from finance and economics, also includes sociophysics and cliophysics.

References

Adams C. (2000). *When in the Course of Human Events: Arguing the Case for the Southern Secession*. Lanham, Maryland: Rowman and Littlefield.

Alesina A., Spolaore E. (2003). *The Size of Nations*. Cambridge, Massachusetts: MIT Press.

Aoki M., Yoshikawa H. (2006). *A Stochastic Approach to Macroeconomics and Financial Markets*. New York: Cambridge University Press.

Asimov I. (1951). *Foundation*. Garden City, New York: Doubleday.

Asimov I. (1952). *Foundation and Empire*. Garden City, New York: Doubleday.

Asimov I. (1953). *Second Foundation*. New York: Gnome Press.

Bakan J. (2004). *The Corporation: The Pathological Pursuit of Profit and Power*. New York: Free Press.

Barabási A.-L. (2002). *Linked: The New Science of Networks*. Cambridge, Massachusetts: Perseus.

Baron-Laforet S. (1999). Repérage du suicide en prison et éléments contextuels. http://psydoc-fr.broca.inserm.fr.

Barsics J., De Rop Y., Lausberg A., Manfroid J. (1993). *Autour du pendule de Foucault*. Liège: Société Astronomique de Liège.

Baudelot C., Establet R. (1984). *Durkheim et le suicide*. Paris: Presses Universitaires de France.

Baudelot C., Establet R. (1985). La sociologie du suicide. *La Recherche* **16**, 162, 12–20.

Bayet A. (1922). *Le suicide et la morale*. Paris: Félix Alcan.

Beales D. (2003). *Prosperity and Plunder: European Catholic Monasteries in the Age of Revolution 1650–1815*. Cambridge: Cambridge University Press.

Behera L., Schweitzer F. (2003). On spatial consensus formation. Is the Sznajd model different from a voter model? *International Journal of Modern Physics C* **14**, 10, 1331–54.

Behr E. (1989). *Hirohito, Behind the Myth*. London: Hamish Hamilton.

Bell G. (1924). Letter from Gertrude Bell to her father, August 13, 1924. http://www.gerty.ncl.ac.uk/letters/11673.htm.

Besnard P. (1976). Anti- ou antédurkheimisme. Contribution au débat sur les statistiques officielles de suicide. *Revue Française de Sociologie*, June, 313–41.

Besnard P. (1997). Mariage et suicide. La théorie durkheimienne de la régulation conjugale à l'épreuve d'un siècle. *Revue Française de Sociologie* **38**, 735–58.

Bidault G. (1965). *D'une résistance à l'autre*. Paris: Les Presses du Siècle.

Bitsch M.-T. (1996). *Histoire de la construction européenne de 1945 à nos jours*. Brussels: Editions Complexes.

243

Blom-Cooper L. (1987). The penalty of imprisonment. Tanner Lectures on Human Values. Delivered at Clare Hall, Cambridge University, November 30–December 2, 1987.

Bodard L. (1972–3). *La guerre d'Indochine*. 5 volumes: *L'enlisement, L'illlusion, L'humiliation, L'aventure, L'épuisement*. Paris: Gallimard.

Bodart G. (1908). *Militär-historisches Kriegs-Lexikon 1618–1905*. Vienna: CW Stern.

Boismont A. B. (1865). *Du suicide et de la folie suicide*. Paris: Germer Baillière.

Bojanovsky J. (1980). Wann droht der Selbstmord bei Verwitweten? [When do suicides of widowers occur?]. *Schweizer Archiv für Neurologie, Neurochirurgie and Psychiatrie* **127**, 1, 99–103.

Bose E. (1907). Resultate kalorimetrischer Studien. *Zeitschrift für Physikalische Chemie* **58**, 585–624.

Botte G. (1911). Le suicide dans l'armée. Etude statistique, étiologique, et prophylactique. Thesis. Lyons.

Boulay R., Lenoir A. (1998). Influence de l'isolement social chez la fourmi *Camponotus fellah*. *Actes des Colloques Insectes Sociaux* **11**, 33–5.

Boulay R., Quagebeur M., Godzinska E. J., Lenoir A. (1999). Social isolation in ants. Evidence of its impact in survivorship and behaviour in *Camponotus fellah* (Hymenoptera Formicae). *Sociobiology* **33**, 111–24.

Boulay R., Soroker V., Godzinska E. J., Hefetz A., Lenoir A. (2000). Octopamine reverses the isolation-induced increase in trophallaxis in the carpenter ant *Camponotus fellah*. *Journal of Experimental Biology* **203**, 513–20.

Bourgoin N. (1999). Le suicide en milieu carcéral. *Population*, May–June, 609–25.

Brocklebank L. (1997). *Jayforce and the Military Occupation of Japan 1945–1948*. Auckland: Oxford University Press.

Bunt T. G. (1851). *Philosophical Magazine* **I**, 4th series, 552.

Campbell P. (1995). *Death in Templecrone*. Jersey City: Princeton Academic Press.

Candiotti C., Dérobert L., Moine C., Moine M. (1948). *Considérations statistiques sur le suicide en France et à l'étranger*. Institut National d'Hygiène.

Cantor C. H., Slater P. J. (1995). Marital breakdown, parenthood, and suicide. *Journal of Family Studies* **1**, 2, 91–102.

CAPAA (Commission on Asian Pacific American Affairs) (2001). *Asian Pacific Heritage Resource Guide*. Seattle.

Carter C. (2002). Between war and peace: the experience of occupation for members of the British Commonwealth Occupation Force, 1945–1952. Thesis submitted to the Australian Defence Academy at the University of New South Wales.

Central Bureau of Statistics of Norway (1978). *Historical Statistics*. Oslo.

Chadwick D. G., Jensen L. (1971). *Water Dowsing: Report of the Utah Water Research Laboratory*. Logan, Utah.

Chang S. (2003). Academic genealogy of American physicists. *AAPPS Bulletin* **13**, 6, 6–41.

Chavanis P.-H., Rosier C., Sire C. (2002). Thermodynamics of self-gravitating systems. *Physical Review E* **66**, 036105-1 to 036105-19.

Chéruy C. editor (2002). *Mortalité après la perte du partenaire: nouvelles données belges* [Mortality after bereavement: new data from Belgium] Brussels: Institut National de la Statistique.

Chesnais J.-C. (1976). *Les morts violentes en France depuis 1826*. Paris: Presses Universitaires de France.

Clarke P. G., Clarke S. (1996). Nineteenth century research on naturally occurring cell death and related phenomena. *Anatomy and Embryology* **193**, 81–99.

Clot A. (1990). *Mehmed II, le conquérant de Byzance*. Paris: Perrin.

Cockett R. (1994). *Thinking the Unthinkable: Think-Tanks and Economic Counter-revolution 1931–1983*. London: HarperCollins.

Coleman P. (1989). *The Liberal Conspiracy: The Congress for Cultural Freedom and the Struggle for the Mind of Postwar Europe*. New York: The Free Press.

Conrad J., Elster L., Lexis W., Loening E. editors (1911). *Handwörterbuch des Staatswissenschaften*, Vol. 7. Jena: Gustav Fischer.

Correctional Service of Canada (1992). Violence and suicide in Canadian institutions: some recent statistics. *Forum* **4**, 3, 1–5.

Corrections Department of New Zealand (2003). Inmate deaths in custody. http://www.corrections.govt.nz.

Cristau C.-A. (1874). *Du suicide dans l'armée*. Thesis. Paris.

Cutlip S. (1994). *The Unseen Power: Public Relations, a History*. Hillsdale, New Jersey: Lawrence Erlbaum Associates.

Daguet F. (1995). *Un siècle de démographie française. Structure et évolution de la population, 1901–1993*. Paris: INSEE.

Dasannacharya B., Heymadi D. (1937). Short Foucault pendulums. *Philosophical Magazine* **23**, 65–88.

DCJ Report. *New York State Division of Criminal Justice Services 1999: Crime and Justice Annual Report* (available at http://criminaljustice.state.ny.us).

Delmas J., Kessler J. editors (1999). *Renseignement et propagande pendant la guerre froide 1947–1953*. Brussels: Editions Complexes.

Demory J.-C. (1995). *Georges Bidault, 1899–1983*. Paris: Julliard.

Depraetere H., Diericks J. (1986). *La guerre froide en Belgique*. Antwerp: Editions EPO.

Deschâtres F., Sornette D. (2005). The dynamics of book sales: endogenous versus exogenous shocks in complex networks. *Physical Review E* **72**, 016112.

Dickens C. (1842, 1996). Philadelphia and its solitary prison. Chapter 7 of *American Notes*. Reedited by Modern Library, New York.

DiLorenzo T. J. (2002). *The Real Lincoln: A New Look at Abraham Lincoln, His Agenda and an Unnecessary War*. New York: Three Rivers Press.

D'Oliveira (1851). *Comptes Rendus de l'Académie des Sciences* **33**, 583.

Donnelly J. S. Jr. (1995). Mass evictions and the Irish famine. In C. Poirteir editor, *The Great Irish Famine*. Dublin: Mercier Press.

Dosal P. (1995). *Power in Transition: The Rise of Guatemala's Industrial Oligarchy 1871–1994*. Westport, Connecticut: Praeger.

Douglas J. D. (1967). *The Social Meanings of Suicide*. Princeton: Princeton University Press.

Dower J. W. (1999). *Embracing Defeat: Japan in the Wake of World War II*. New York: W. W. Norton.

Drame S., Gonfalone C., Miller J. A., Roehner B. (1991). *Un siècle de commerce du blé en France, 1825–1913*. Paris: Economica.

Dubois P. (2003). Culture et guerre froide 1945–1953. *Relations Internationales* **115**, Autumn, 437–54.

Dufour, Wartman, Marignac (1851). *Comptes Rendus de l'Académie des Sciences* **33**, 13.

Durkheim E. (1897). *Le suicide: étude de sociologie*. Paris: Félix Alcan.

Duverger C. (2001). *Cortes*. Paris: Fayard.

Eichelberger R. L. *General Eichelberger's papers*. Edited on microfilm by Adam Matthew Publications, London.

Engel E. (1861). Die Getreidepreise, die Ernteerträge und der Getreidehandel im preussischen Staate [Grain prices, yields and trade in Prussia]. *Zeitschrift des Königliche Preussischen Statistischen Bureaus* **10**, 11249–89.

Eurostat (2004). *Statistiques de population*. Chapter G: Nuptialité [marriages], Brussels.

Ewen S. (1996). *PR! A Social History of Spin*. New York: Basic Books.

Finn R. B. (1992). *Winners in Peace: MacArthur, Yoshida and Postwar Japan*. Berkeley: University of California Press.

Flamigni S. (2004). *La sfinge delle Brigate Rosse* [The sphinx of the Red Brigades]. Milan: Kaos Edizioni.

Fleury G. (1994). *La guerre en Indochine*. Paris: Plon.

Flora P., Alber J., Eichenberg R., *et al.* (1983). *State, Economy, and Society in Western Europe 1815–1975: A Data Handbook in Two Volumes*. Vol. 1: *The Growth of Mass Democracies and Welfare States*. London: Macmillan.

Flora P., Kraus F., Pfenning W. (1987). *State, Economy, and Society in Western Europe 1815–1975: A Data Handbook in Two Volumes*. Vol. 2: *The Growth of Industrial Societies and Capitalist Economies*. London: Macmillan.

Foucault L. (1851). *Comptes Rendus de l'Académie des Sciences* **32**, 135.

Foucault L. (1878). *Recueil des travaux scientifiques de Léon Foucault*. Published by Léon Foucault's mother. Paris: Gauthier-Villars.

Fry R. (2005). *The Higher Dropout Rate of Foreign-Born Teens: The Role of Schooling Abroad*. Pew Hispanic Center, Report of November 1, 2005.

Fujioka T. (1978). Study of ice growth in slightly undercooled water. Doctoral Thesis, Department of Metallurgy and Material Science, Carnegie Mellon University, Pittsburgh.

Gaskill M. (2005). *Witchfinders. A Sevententh-Century English Tragedy*. London: John Murray.

Geiger H., Marsden E. (1909). On the diffuse reflection of the α-particles. *Proceedings of the Royal Society A* **82** (May), 495–500.

Goetzmann W. N., Ukhov A. (2001). *China and the World Financial Markets 1870–1930: Modern Lessons from Historical Globalization*. Philadelphia: Wharton School publishing, University of Pennsylvania.

Goodrich T. (1995). *Black Flag: Guerilla Warfare on the Western Border 1861–1865*. Bloomington: Indiana University Press.

Grafton-Cardwell B. (2005). Overview of 2005 pest management. *Citrus Notes* **2**, 2, 2 (Tulare, California).

Grassé P.-P., Chauvin R. (1944). L'effet de groupe dans la survie des neutres dans les sociétés d'insectes [Collective effects in the survival of drones in insect societies]. *La Revue Scientifique* Oct.–Nov., 461–4.

Grove R. D., Hetzel A. M. (1968). *Vital Statistics in the United States 1940–1960*. Washington, DC: National Center of Health Statistics.

Grubel H. G. (1998). Economic freedom and human welfare: some empirical findings. *Cato Journal* **18**, 2, 287–304.

Guiffan J. (1989). *La question d'Irlande*. Paris: Editions Complexes.

Halbwachs M. (1930). *Les causes du suicide*. Paris: Félix Alcan.

Hartenian L. R. (1984). Propaganda and the control of information in occupied Germany: the U.S. information control division of radio Frankfurt 1945–1949. Thesis, State University of New Jersey, Rutgers.

Hart-Landsberg M. (2002). Challenging neoliberal myths: a critical look at the Mexican experience. *Monthly Review* **54**, 7.

Hayek F. A. (1944). *The Road to Serfdom*. Chicago: University of Chicago Press.

Hayek F. A. (1994). *Hayek on Hayek: An Autobiographical Dialogue*. Chicago: University of Chicago Press.

Hayes L. M., Rowan J. R. (1988). *National Study of Jail Suicides: Seven Years Later*. Alexandria, Virginia: National Center on Institutions and Alternatives.

Hezel F. X. (1987). Truk [called Chuuk after 1990] suicide epidemic and social change. *Human Organization* **48**, 283–96.

Hezel F. X. (2001). Competition drives school improvement. *Pacific Magazine and Island Business* (May).

HMPS (Her Majesty's Prison Service) (2001). *Prison Suicides*. London.

Holmes J. (1913). Contributions to the theory of solution: the intermiscibility of liquids. *Journal of the Chemical Society* 2147–66.

Hyer G., Lund E. (1993). Suicide among women related to number of children in marriage. *Archive General of Psychiatry* **50**, 134–6.

Ishizaki Y., Burne J. F., Raff M. C. (1994). Autocrine signals enable chondrocytes to survive in culture. *The Journal of Cell Biology* **126**, 1069–77.

Kamerlingh Onnes H. (1879). Nieuwe bewijzen voor de aswenteling des aarde [New proofs of the rotation of the earth] Thesis. Groningen.

Kariya Y., Gagg C., Plumbridge W. J. (2000). The pest in lead free [i.e. pure tin] solders. *Soldering and Surface Mount Technology* **13**, 1, 39–40.

Keddie N. R. (1981). *Roots of Revolution. An Interpretative History of Modern Iran.* New Haven: Yale University Press.

Kelly S., Bunting J. (1998). *Trends in Suicides in England and Wales 1982–1996.* London: Office of National Statistics.

Knightley P. (2004). *The First Casualty: The War Correspondent as Hero and Myth-Maker from the Crimea to Iraq.* Baltimore: Johns Hopkins University Press.

Krose H. A. (1906). *Der Selbstmord im 19. Jahrhundert.* Fribourg: Herdersche Verlag.

Landolt-Börnstein edited by K. Schäfer (1976). *Numerical Data and Functional Relationships in Science and Technology. Group IV, Vol. 2: Heats of Mixing and Solution.* Berlin: Springer.

Langer W. L. editor (1968). *An Encyclopedia of World History.* Boston: Houghton Mifflin Company.

Langley J., Langley S. (1989). *State-Level Wheat Statistics 1949–1988.* US Department of Agriculture, Statistical Bulletin No. 779.

Lecky W. E. H. (1892). *A History of England in the 18th Century.* 8 volumes. London: Longmans and Green.

Legoyt A. (1881). *Le suicide ancien et moderne: étude historique, philosophique et statistique.* Paris: A. Droin.

Le Roy E. (1870). *Etude sur le suicide et les maladies mentales dans le département de Seine-et-Marne.* Paris: Victor Masson.

Lester D. (1997). *Suicide in American Indians.* New York: Nova Science.

Lever E. (1996). *Philippe Egalité.* Paris: Fayard.

Lide D. R. editor (2001). *CRC Handbook of Chemistry and Physics.* Cleveland: CRC Press.

Lieberson S. (1980). *A Piece of the Pie: Blacks and White Immigrants Since 1880.* Berkeley: University of California Press.

Lieberson S. (1985). *Making It Count: The Improvement of Social Research and Theory.* Berkeley: University of California Press.

Liesner T. (1989). *One Hundred Years of Economic Statistics.* New York: Facts on File.

Linder F. E., Grove R. D. (1947). *Vital Statistics Rates in the United States 1900–1940.* Washington, DC: Government Printing Office.

Lucier G., Budge A., Plummer C., Spurgeon C. (1991). *U.S. Potato Statistics.* US Department of Agriculture. Statistical Bulletin No. 829.

Ludwig B. (2003). La progagande anticommuniste en Allemagne Fédérale. Vingtième Siècle. *Revue d'Histoire* **80**, Oct–Dec, 33–42.

Luoma J. B., Pearson J. L. (2002). Suicide and marital status in the United States 1991–1996. Is widowhood a risk factor? *American Journal of Public Health* **92**, 9, 1518–22.

MacDonald M., Sexton S. (2002). *Self-Harm and Suicide Policy: Implementation in West Midlands Prisons*. Birmingham: University of Central England Report.

Malinvaud E. (1990). Propos de circonstances sur les orientations de la discipline économique. *Annales, Economie, Société, Civilisation* **1**, 115–21.

Manchester W. R. (1978). *American Caesar, Douglas MacArthur 1890–1964*. Boston: Little, Brown.

Mandelbrot B. B. (1975). *Les objets fractals: forme, hasard et dimension*. Paris: Flammarion.

Mantegna R. N., Stanley H. E. (2000). *An Introduction to Econophysics: Correlations and Complexity in Finance*. Cambridge: Cambridge University Press.

Marshall T., Simpson S., Stevens A. (2000). *Health Care in Prisons*. University of Birmingham Report (February 2000).

Masaryck T. G. (1881). *Der Selbstmord als soziale Massenerscheinung der modernen Civilisation*. Vienna: Carl Konegen.

Maslov S., Sneppen K. (2002). Specificity and stability of protein networks. *Science* **296**, 910–13.

Mason R. H. P., Caiger J. G. (1973, 1997). *A History of Japan*. Boston: Tuttle Publishing.

Matsumoto K. K. (1975). Influence of folk superstition on fertility of Japanese in Hawaii (1966). *American Journal of Public Health* **65**, 2, 170–4.

May R. (1976). Simple mathematical models with very complicated dynamics. *Nature* **261**, 459–67.

McInnes N. (1998). The road not taken: Hayek's slippery road to serfdom. *The National Interest*, Spring.

Meer F. (1976). *Race and Suicide in South Africa*. London: Routledge and Kegan Paul.

Milgram S. (1974, 1993). *Obedience to Authority: An Experimental View*. Reading, UK: Pinter and Martin.

Mingay G. E. (1956). Estate management in eighteenth century Kent. *The Agricultural History Review* **4**, 108–13.

Mitchell B. R., Deane P. (1971). *Abstract of British Historical Statistics*. Cambridge: Cambridge University Press.

Monographies de produits 1950: Céréales. Washington, DC: Organisation des Nations Unies pour l'alimentation et l'agriculture.

Morgenstern O. (1950). *On the Accuracy of Economic Observations*. Princeton: Princeton University Press.

Moriguchi C., Saez E. (2004). The evolution of income concentration in Japan 1885–2002. http://emlab.berkeley.edu/users/saez/moriguchi-saez05japan.pdf

Morselli E. (1879). *Il suicidio, saggio di statistica morale comparata*. Milan: Fratelli Dumolard.

Nagle J. T. (1882). *Suicides in New York City During the 11 Years Ending December 31, 1880*. Cambridge, Massachusetts: Riverside Press.

Nearing S., Freeman J. (1926). *Dollar Diplomacy: A Study in American Imperialism*. London: George Allen and Unwin.

Nizard A. (1998). Suicide et mal-être social. *Population et Sociétés* **334**, 2–12.

Nolt S. M. (1992). *A History of the Amish*. Intercourse, Pennsylvania: Good Books.

Nordyke E. C. (1989). *The Peopling of Hawaii*. Honolulu: University of Hawaii Press.

Nozdrev V. F. (1965). *The Use of Ultrasonics in Molecular Physics*. Oxford: Pergamon Press.

Olson L. M. (2003). *Suicide in American Indians*. University of Utah.

Onnes K. H. (1879). Nieuwe bewijzen voor de aswenteling des aarde [New proofs of the rotation of the earth] Thesis. Groningen.

Papoulis A. (1965). *Probability, Random Variables and Stochastic Processes*. Tokyo: McGraw-Hill.

Penrose R. (1989). *The Emperor's New Mind*. New York: Oxford University Press.

Peterson E. N. (1990). *The Many Faces of Defeat: The German People's Experience in 1945*. American University Studies Number 9. New York: Peter Lang.

Pfeiffer M. B. (2001). Suicides high in prison box. *Poughkeepsie Journal* [New York State], December 16, 2001.

Pfeiffer M. B. (2002). Suicides in solitary confinement are abnormaly high. *Poughkeepsie Journal* [New York State], April 14, 2002.

Phillips D. P. (1974). The influence of suggestion on suicide: substantive and theoretical implications of the Werther effect. *American Sociological Review* **39**, 340–54.

Phillips J. (1851). *Philosophical Magazine*. **II**, 4th series,150.

Piketty T., Saez E. (2003). Income inequality in the United States 1913–1998. *Quarterly Journal of Economics* **118**, 1, 1–39.

Poirteir C. (1995). *Famine Echoes*. Dublin: Gill and Macmillan.

Puentes-Rosas E., López-Nieto L., Martinez-Monroy T. (2004). La mortalidad por suicidios: México 1990–2001. *Pan American Journal of Public Health* **16**, 2, 102–9.

Quentin M., Bouchaud J.-P. (2005). Theory of collective opinion shifts: from smooth trends to abrupt swings. Cond-mat/0504079 (4 April).

Quid (1997). Edited by D. Frémy and M. Frémy. Paris: Robert Laffont.

Raff M. C. (1998). Cell suicide for beginners. *Nature* **396**, 12 November, 119–22.

Rampton S., Stauber J. (2001). *Trust Us, We're Experts*. New York: Penguin Putnam.

Reif F. (1965). *Fundamentals of Statistical and Thermal Physics*. New York: McGraw-Hill.

Rival M. (1995). *Robert Oppenheimer*. Paris: Flammarion.

Rocard Y. (1981). *Les sourciers (Que sais-je? No. 1939)*. Paris: Presses Universitaire de France.

Rodnesky S. (2004). *Conning the Rich: The Great American Fraud*. Pembroke Pines, Florida: BMES Press.

Roehner B. M. (1995). *Theory of Markets*. Berlin: Springer.

Roehner B. M. (2000a). The correlation length of commodity markets. 1 Empirical evidence. *European Physical Journal B*, **13**, 175–87.

Roehner B. M. (2000b). The correlation length of commodity markets. 2 Theoretical framework. *European Physical Journal B*, **13**, 189–200.

Roehner B. M. (2001). *Hidden Collective Factors in Speculative Trading*. Berlin: Springer.

Roehner B. M. (2002). *Patterns of Speculation: A Study in Observational Econophysics*. Cambridge: Cambridge University Press.

Roehner B. (2004). *Cohésion sociale*. Paris: Odile Jacob.

Roehner B. M. (2005a). A bridge between liquids and socio-economic systems: the key role of interaction strengths. *Physica A* **348**, 659–82.

Roehner B. M. (2005b). Macro-players in stock markets. In *Practical Fruits of Econophysics*, H. Takayasu editor. Tokyo: Springer.

Roehner B. M. (2005c). Transfat: risk, profit and consensus formation models. *Proceedings of the Viterbo Conference on Risk Management*, 2 September 2005.

Roehner B. M., Rahilly L. J. (2002). *Separatism and Integration*. Lanham, Maryland: Rowman and Littlefield.

Roehner B. M., Sornette D., Anderson J. V. (2004). Response functions to critical shocks in social sciences: an empirical and numerical study. *International Journal of Modern Physics C*, **15**, 6, 809–34.

Roehner B. M., Syme T. (2002). *Patterns and Repertoire in History*. Cambridge, Massachusetts: Harvard University Press.

Ropp P. S., Zamperini P., Zurndorfer H. T. editors (2001). *Passionate Women: Female Suicide in Late Imperial China*. Boston: Brill.

ROSPA (The Royal Society for the Prevention of Accidents) (2002). *The Risk of Using a Mobile Phone While Driving*.

Roussel E. (1996). *Jean Monnet, 1888–1979*. Paris: Fayard.

Rubinstein D. H. (2002). *Youth Suicide and Social Change in Micronesia*. Occasional Paper, Kagoshima University, 36, 33–41.

Rutherford E. (1911). The scattering of alpha and beta particles by matter and the structure of the atom. *Philosophical Magazine* **21**, 669–88.

Salisbury H. E. (1980). *Without Fear or Favor: The New York Times and Its Times*. New York: Quadrangle Times Books.

Sattar G. (2001). Rates and causes of death among prisoners and offenders under community supervision. Home Office Research Study 231 [available on the website of the Home Office www.homeoffice.gov.uk].

Schimdtke A. *et al.* [26 co-authors] (1999). Suicide rates in the world: update. *Archives of Suicide Research* **5**, 81–9.

Schulz-Dubois E. O. (1970). Foucault pendulum experiment by Kamerlingh Onnes and degenerate perturbation theory. *American Journal of Physics* **38**, 2, 173–88.

Scoville J. W., Sargent N. editors (1942). *Facts and Fancy in the T.N.E.C. Monographs*. Sponsored by the National Association of Manufacturers, New York.

Seuffert G. K. L. (1857). *Statistik des Getreide und Viktualien-Handels im Königreiche Bayern*. Munich: Weiß.

Sevela M. editor (2001). Memoirs of D. N. Kriukov on the civil administration on South Sakhalin and the Kurile Islands, 1945–1948. *Monumenta Nipponica* **56**, 1.

Smith J. C., Mercy J. A., Conn J. M. (1988). Marital status and the risk of suicide. *American Journal of Public Health* **78**, 1, 78–80.

Sornette D. (2003). *Why Stock Markets Crash: Critical Events in Complex Financial Analysis*. Princeton: Princeton University Press.

Sornette D., Deschâtres F., Gilbert T., Ageon Y. (2004). Endogenous versus exogenous shocks in complex networks: an empirical test using book sales rankings. *Physical Review Letters* **93**, 228701–4.

Stack S. (1987). Celebrities and suicide. A taxonomy and analysis, 1948–1983. *American Sociological Review* **52**, 401–12.

Stack S. (1989). The impact of suicide in Norway, 1951–1980. *Journal of Marriage and the Family* **51**, 229–38.

Stack S. (1990). The effects of divorce on suicide in Denmark, 1961–1980. *The Sociological Quarterly* **31**, 361–8.

Staley E. (1935). *War and the Private Investor: A Study in the Relations of International Politics and International Private Investment*. Garden City, New York: Doubleday, Doran and Co.

Statistiske Centralbyrå (1926). *Folkemengdens bevegelse 1911–1920*. Oslo: H. Aschehoug.

Stauffer D. (2003). How to convince others? Monte Carlo simulations of the Sznajd model. In *The Monte Carlo Method in the Physical Sciences*, J. Gubernatis editor. AIP Conference Proceedings.

Steenkamp M., Harrison J. (2000). *Suicide and Hospitalized Self-Harm in Australia.* Australian Institute of Health and Welfare.

Strogatz S. H. (2003). *Sync: The Emerging Science of Spontaneous Order.* New York: Hyperion.

Sundeen M. (2004). Cell phones and highway safety: 2003 state legislative update. National Conference of State Legislatures (January 23, 2004). www.ncsl.org/programs/transportation/cellphoneupdate1203.htm.

Telzrow M. E. (2002). Punishment and reform: the Wisconsin State reformatory. *Voyageur*, Winter–Spring, p. 25.

Tocqueville A. (1835, 1958). *Journeys to England and Ireland.* London: Faber and Faber.

Tocqueville A. (1991). *Oeuvres complètes.* Paris: Gallimard.

Treat P. J. (1928). *The Far East: A Political and Diplomatic History.* New York: Harper and Brothers.

Tsurumi S. (1961). *Haikyo no naka kara 1945–1952* [Emerging from the ruins, in Japanese]. Tokyo: Chikuma Shobo.

Turchin P. (2003a). *Complex Population Dynamics: A Theoretical/Empirical Synthesis.* Princeton: Princeton University Press.

Turchin P. (2003b). *Historical Dynamics: Why States Rise and Fall.* Princeton: Princeton University Press.

Ulrich R. (1944). La chute des fruits. *La Revue Scientifique* **82**, January, 25–34.

Ventsel H. (1973). *Théorie des probabilités.* Moscow: Mir.

Villermé L. R (1820). *Des prisons, telles qu'elles sont et telles qu'elles devraient être.* Paris: Méquignon-Marvis.

Wade J. (1832, 1970). *The Extraordinary Black Book: An Exposition of Abuses in Church and State, Courts of Law, Representation, Municipal and Corporate Bodies with a Precis of the House of Commons.* New York: Augustus M. Kelly.

Walpen B., Plehwe D. (2001). Wahrheitsgetreue Berichte über Chile – Die Mont Pélerin Society und die Diktatur Pinochet. *1999 Zeitschrift für Sozialgeschichte des 20. und 21. Jahrhunderts* **16**, 2 (September 2001).

Watts D. (2003). *Six Degrees: The Science of a Connected Age.* London: Norton.

Weyerer S., Wiedenmann A. (1995). Economic factors and the rates of suicide in Germany between 1881 and 1989. *Psychological Reports* **76**, 1331–41.

Zacharachis C. A., Madianos M. G., Papadimitriou G. N., Stephanis C. N. (2003). Epidemiology of suicide in Greece 1980–1997. *Archives of Hellenic Medicine* **20**, 2, 191–9.

Zorgbibe C. (1978). *La construction politique de l'Europe 1946–1976.* Paris: Presses Universitaires de France.

Zorgbibe C. (2005). *Histoire de l'Union Européenne.* Paris: Albin Michel.

Index